BROOKINGS INSTITUTION, WASHINGTON, D.C.
INSTITUTE FOR GOVERNMENT RESEARCH

SERVICE MONOGRAPHS
OF THE
UNITED STATES GOVERNMENT
No. 58

THE FOREST SERVICE

ITS HISTORY, ACTIVITIES AND ORGANIZATION

AMS PRESS

NEW YORK

THE BROOKINGS INSTITUTION

The Brookings Institution—Devoted to Public Service through Research and Training in the Humanistic Sciences—was incorporated on December 8, 1927. Broadly stated, the Institution has two primary purposes: the first is to aid constructively in the development of sound national policies; and the second is to offer training of a super-graduate character to students of the social sciences. The Institution will maintain a series of coöperating institutes, equipped to carry out comprehensive and inter-related research projects.

The responsibility for the final determination of the Institution's policies and for the administration of its endowment is vested in a self-perpetuating Board of Trustees. The Trustees have, however, defined their position with reference to the scientific work of the Institution in a by-law provision reading as follows: "The primary function of the Trustees is not to express their views upon the scientific investigations conducted by any division of the Institution, but only to make it possible for such scientific work to be done under the most favorable auspices." Major responsibility for "formulating general policies and coördinating the activities of the various divisions of the Institution" is vested in the President. The by-laws provide also that "there shall be an Advisory Council selected by the President from among the scientific staff of the Institution and representing the different divisions of the Institution."

INSTITUTE FOR GOVERNMENT RESEARCH

SERVICE MONOGRAPHS
OF THE
UNITED STATES GOVERNMENT
No. 58

THE FOREST SERVICE

ITS HISTORY, ACTIVITIES
AND ORGANIZATION

BY

DARRELL HEVENOR SMITH

THE BROOKINGS INSTITUTION
WASHINGTON
1930

Library of Congress Cataloging in Publication Data

Smith, Darrell Hevenor.
  The Forest Service.

    Original ed. issued as no. 58 of Service monographs
  of the United States Government.
    Bibliography: p.
    1. United States. Forest Service. I. Title. II. Series:
  Brookings Institution, Washington D.C.
  Institute for Government Research. Service monographs
  of the United States Government, no. 58.
  SD565.S6   1974   353.008'233   72-3075
  ISBN 0-404-57158-1

Reprinted from the edition of 1930, Washington, D.C.
First AMS edition published, 1974
Manufactured in the United States of America

International Standard Book Number:
Complete Set: 0-404-57100-X
Volume 58:   0-404-57158-1

AMS Press, Inc.
New York, N.Y. 10003

FOREWORD

The first essential to efficient administration of any enterprise is full knowledge of its present make-up and operation. Without full and complete information before them, as to existing organization, personnel, plant, and methods of operation and control, neither legislators nor administrators can properly perform their functions.

The greater the work, the more varied the activities engaged in, and the more complex the organization employed, the more imperative becomes the necessity that this information shall be available—and available in such a form that it can readily be utilized.

Of all undertakings, none in the United States, and few, if any, in the world, approach in magnitude, complexity, and importance that of the national government of the United States. As President Taft expressed it in his message to Congress of January 17, 1912, in referring to the inquiry being made under his direction into the efficiency and economy of the methods of prosecuting public business, the activities of the national government " are almost as varied as those of the entire business world. The operations of the government affect the interest of every person living within the jurisdiction of the United States. Its organization embraces stations and centers of work located in every city and in many local subdivisions of the country. Its gross expenditures amount to billions annually. Including the personnel of the military and naval establishments, more than half a million persons are required to do the work imposed by law upon the executive branch of the government.

" This vast organization has never been studied in detail as one piece of administrative mechanism. Never have the foundations been laid for a thorough consideration of the relations of all of its parts. No comprehensive effort has been made to list its multifarious activities or to group them in such a way as to present a clear picture of what the government is doing. Never has a complete description been given of the agencies through which these activities are performed. At no time has the attempt been made to study all of these activities and agencies with a view to the assignment of each activity to the agency best fitted for its performance, to the avoidance of duplication of plant and work, to the integration of all administrative agencies of the government, so far as may be practicable, into a unified organization for the most effective and economical dispatch of public business."

To lay the basis for such a comprehensive study of the organization and operations of the national government as President Taft outlined, the Institute for Government Research has undertaken the preparation of a series of monographs, of which the present study is one, giving a detailed description of each of the distinct services of the government. These studies are being vigorously prosecuted, and it is hoped that all services of the government will be covered in a comparatively brief space of time. Thereafter, revisions of the monographs will be made from time to time as need arises, to the end that they may, as far as practicable, represent current conditions.

These monographs are all prepared according to a uniform plan. They give: first, the history of the establishment and development of the service; second, its functions, described not in general terms, but by detailing its specific activities; third, its organization for the handling of these activities; fourth, the character of its plant; fifth, a compilation of, or reference to, the laws and regulations governing its operations; sixth, financial statements showing its appropriations, expenditures and other data for a period of years; and finally, a full bibliography of the sources of information, official and private, bearing on the service and its operations.

In the preparation of these monographs the Institute has kept steadily in mind the aim to produce documents that will be of direct value and assistance in the administration of public affairs. To executive officials they offer valuable tools of administration. Through them, such officers can, with a minimum of effort, inform themselves regarding the details, not only of their own services, but of others with whose facilities, activities, and methods it is desirable that they should be familiar. Under present conditions services frequently engage in activities in ignorance of the fact that the work projected has already been done, or is in process of execution by other services. Many cases exist where one service could make effective use of the organization, plant or results of other services had they knowledge that such facilities were in existence. With the constant shifting of directing personnel that takes place in the administrative branch of the national government, the existence of means by which incoming officials may thus readily secure information regarding their own and other services is a matter of great importance.

To members of Congress the monograph should prove of no less value. At present these officials are called upon to legislate and appropriate money for services concerning whose needs and real problems they can secure but imperfect information. That the possession by each member of a set of monographs such as is here projected, prepared according to a uniform plan, will be a great aid to intelligent legislation and appropriation of funds can hardly be questioned.

To the public, finally, these monographs will give that knowledge of the organization and operations of their government which must be had if an enlightened public opinion is to be brought to bear upon the conduct of governmental affairs.

These studies are wholly descriptive in character. No attempt is made in them to subject the conditions described to criticism, nor to indicate features in respect to which changes might with advantage be made. Upon administrators themselves falls responsibility for making or proposing changes which will result in the improvement of methods of administration. The primary aim of outside agencies should be to emphasize this responsibility and facilitate its fulfillment.

While the monographs thus make no direct recommendations for improvement, they cannot fail greatly to stimulate efforts in that direction. Prepared as they are according to a uniform plan, and setting forth as they do the activities, plant, organization, personnel and laws governing the several services of the government, they will automatically, as it were, reveal, for example, the extent to which work in the same field is being performed by different services, and thus furnish the information that is essential to a consideration of the great question of the better distribution and coördination of activities among the several departments, establishments, and bureaus, and the elimination of duplication of plant, organization and work. Through them it will also be possible to subject any particular feature of the administrative work of the government to exhaustive study, to determine, for example, what facilities, in the way of laboratories and other plant and equipment, exist for the prosecution of any line of work and where those facilities are located; or what work is being done in any field of administration or research, such as the promotion, protection and regulation of the maritime interests of the country, the planning and execution of works of an engineering character, or the collection, compilation and publication of statistical data, or what differences of practice prevail in respect to organization, classification, appointment, and promotion of personnel.

To recapitulate, the monographs will serve the double purpose of furnishing an essential tool for efficient legislation, administration and popular control, and of laying the basis for critical and constructive work on the part of those upon whom responsibility for such work primarily rests.

CONTENTS

THE FOREST SERVICE

ITS HISTORY, ACTIVITIES, AND ORGANIZATION

CHAPTER I

HISTORY

The Forest Service is a bureau of the Department of Agriculture. It administers the national forests and seeks to promote the best use of all forests and forest products of the United States through research, coöperation with states and with private agencies, and the diffusion of useful information on forestry. It traces its history uninterruptedly from 1876, in which year a forestry agency was set up under the Commissioner of Agriculture. By 1881 this agency had developed into a Division of Forestry, which, in 1886, received statutory recognition (Act of June 30, 1886; 24 Stat. L., 100, 103). It became the Bureau of Forestry in 1901 (Act of March 2, 1901; 31 Stat. L., 922, 929).[1]

Since July 1, 1905, this bureau has been known as the Forest Service, but the change in designation effected no change in formal status.

Early Efforts at Protection.[2] The year 1776 found the Atlantic seaboard and a thin fringe of the back country carved out of the forest occupied by a new nation, economically based, in no small degree, upon wealth created out of the forest, and fac-

[1] In the Annual Report of the Commissioner of Agriculture for 1884, and in Volume IV of the "Report upon Forestry," published the same year, reference is made to a "Bureau of Forestry," but in 1883 and from 1885 to 1900 the term "division" is used.

[2] For a more detailed statement, see Cameron, Development of Governmental Forest Control in the United States (1928). Institute for Government Research, Studies in Administration.

ing a continent containing forest resources awaiting development compared to which those that had been utilized since 1607 were insignificant.

To the cursory glance of that time American forest resources might well have seemed unlimited, but accessibility had to be taken into consideration. Evidence is available to the effect that scarcely had the colonists begun to establish themselves before they were confronted with the danger of local timber shortage and they legislated to prevent it. Such shortages concerned fuel principally, but by 1799 the possibility of a general shortage, involving ship-construction timber, presented itself. The actions resulting from these two phases of the timber problem differed widely.

"Live-oak" Legislation. The possibility of a general shortage of ship-construction timber (principally live oak) for the navy elicited early action. The source of supply was limited to the southeastern section of the country. On February 25, 1799, an act was passed (1 Stat. L., 622) " authorizing the purchase of timber for naval purposes." This was followed, eighteen years later (March 1, 1817) by an act " making reservation of certain public lands to supply timber for naval purposes " (3 Stat. L., 347), which was more comprehensive in its provisions and more severe in punishments for violations.

On February 23, 1822, came an act " for the preservation of the timber of the United States in Florida " (3 Stat. L., 651), which went a step further and authorized the President to employ the military forces of the United States to prevent the destruction and theft of government timber. This was followed by "An act for the gradual improvement of the Navy of the United States," approved March 3, 1827 (4 Stat. L., 242), which made further timber conservation possible by providing:

That the President of the United States be, and he is hereby, authorized to take the proper measures to preserve the live oak timber growing on the lands of the United States, and he is also authorized to reserve from sale such lands, belonging to the

United States, as may be found to contain live oak, or other timber in sufficient quantity to render the same valuable for naval purposes.[3]

Meanwhile, in 1821, an early act, which made no direct reference to timber, was construed by the Attorney General as applying to the forests. This was the Anti-Trespass Act of March 3, 1807 (2 Stat. L., 445), "To prevent settlements being made on lands ceded to the United States, until authorized by law," under which timber stealers could now be removed from public lands by military force. This applied to all public forest land.[4]

The Attorney General held that stealing public timber in general, as contrasted with the theft of naval live oak, merited, "from its greater frequency and the greater secrecy and security with which it may be committed . . . a still severer reprehension." This opinion was transmitted to all registers and receivers of land offices. These "live-oak acts" and the opinion just mentioned constitute the earliest efforts toward the formulation of a national policy with regard to the reservation, protection, and conservation of timber on the public domain.

Protection. A further group of laws, intimately related to the live-oak acts, deserves mention at this point: Laws designed for the protection of reservations, and, by construction if not wording, the public forests generally.

One such law was approved March 2, 1831 (4 Stat. L., 472): "To provide for the punishment of offenses committed in cutting, destroying or removing live oak and other timber or trees reserved for naval purposes." This act constitutes the basic present-day law for the prevention of timber trespass on government land, for while it was designed primarily to cover naval timber on live-oak reservations, the language was broad enough to cover all timber on the public lands of the United States.

[3] This provision was reinforced by the act of March 19, 1828 (4 Stat. L., 254, 256).

[4] 1 Op. Atty. Gen., 471. Confirmed by the Supreme Court in United States v. Briggs, 9 Howard, 356 (1850).

The act of March 2, 1833 (4 Stat. L., 646, 647), provided for additional inspection of timber shipped by sea, in order to prevent the exportation of live oak.

While the results of these acts were not all that could be desired, numerous concrete accomplishments are to be recorded. Between the beginning of the nineteenth century and the Civil War a total of over 264,000 acres of live-oak land in the southern coastal states was set aside from the public lands and some two thousand additional acres were purchased from private owners.[5]

Enforcement of the law against timber trespass on the public domain was at first placed under the Treasury Department, which acted through " timber agents." In 1854 this responsibility was transferred to the Interior Department, which acted through the General Land Office. During the fifties measures to curb timber depredations were applied in the pine forests of Michigan, Wisconsin, and Minnesota, and on December 24, 1855, the Commissioner of the General Land Office issued a circular making registrars and receivers of land offices responsible for the protection of public timber within their districts. These officers appointed special deputies for the prevention of timber trespass.

The sundry civil appropriation act of June 10, 1872 (17 Stat. L., 347, 359), contained an item of ten thousand dollars " for suppressing depredations on the public timber," through the activities of the special deputies. This was the first direct appropriation for the protection of public timber in general, and included timber of all types.

Tree Planting. The measures just discussed looked to the problems of prevention and protection, but meanwhile a more constructive movement had arisen; namely, " tree planting." As early as 1830 certain residents of Missouri had petitioned Congress to grant them a township of prairie land " for the purpose of making an experiment of raising of forest timber ";[6] and

---

[5] 40 Cong. 2 sess., H. doc. 161.
[6] The grant was denied. See 21 Cong. 1 sess., S. doc. 127.

from time to time prudent men gave expression to opinions on the iniquity of timber waste and to misgivings on the matter of future supply. In 1858, the Georgia legislature asked Congress to appoint a commission " to inquire into the limits, extent and probable duration of the southern pine belt." [7] In 1860, the Agricultural Division of the Patent Office devoted thirty pages of its annual report to a consideration of the " Forest Trees of North America." During the ensuing decade the tree-planting idea became a popular movement in the prairie states. In 1866 the Commissioner of Agriculture devoted two pages of his annual report to a discussion of the promotion of artificial forestation in treeless regions.

Michigan and Wisconsin in 1867 inaugurated inquiries into their forest conditions and needs. Local legislation encouraged tree growing by bounties and tax exemptions. [8] The Board of Agriculture of Maine appointed a committee in 1869 to report on a forestry policy for the state, and three years later a law was passed " for the encouragement of the growth of trees," which exempted from taxation, for twenty years, all land planted to trees.

On April 10, 1872, Nebraska, the " Tree Planters' State " launched the "Arbor Day" idea. During this same year the Commissioner of Agriculture gave his approval to the recommendation of tree planting made by his predecessor in 1866, [9] while the state of New York created a commission to consider state ownership of "the wild lands lying northward of the Mohawk, or so much thereof as may be expedient."

The national government now took action and on the third of March, 1873, the " Timber-Culture Law " (17 Stat. L., 605) was approved, " to encourage the growth of timber on Western prairies." This was to be done through the encouragement of artificial forestation by the same general means by

[7] 35 Cong. 2 sess., S. misc. doc. 12.

[8] Between 1868 and 1872 such laws were passed by Connecticut, New York, Minnesota, Wisconsin, Iowa, Missouri, Dakota, Nebraska, Kansas, and Nevada.

[9] General Land Office, Annual Report, 1872, p. 69.

2

which agriculture was stimulated under the Homestead Act; that is, by requiring the planting and successful growing of a certain number of timber trees as the consideration of a deed to a quarter-section of the public domain.[10]

It is difficult to trace the development of thought and activity on the various phases of forest protection, preservation, and replenishment prior to 1873, and still more difficult to appraise the part each played in advancement toward true forest administration in America. It is clear, however, that the seventies witnessed concrete accomplishments, and one event of those years stands out as having had immediate results. This was the annual meeting of the American Association for the Advancement of Science, held at Portland, Maine, in August, 1873. From this point onward the progress of governmental forestry may be accurately followed.

The Beginnings of Governmental Forestry. At the Portland meeting one of the members, Dr. Franklin B. Hough of Lowville, New York, presented a paper on "The Duty of Governments in the Preservation of Forests." This resulted in the appointment of a committee to memorialize Congress and the state legislatures on the urgent need for governmental action with regard to forest protection and the cultivation of timber. Legislation to the latter end was recommended and a subcommittee was appointed to prepare the memorial and further its consideration.[11] This memorial, together with the draft of a suggested joint resolution of Congress embodying the ideas of the Association as to necessary legislation, was transmitted to Congress with a special message by President Grant on Feb-

[10] The act was a failure in spite of numerous amendments (18 Stat. L., 21; 19 Stat. L., 54; 19 Stat. L., 59; 20 Stat. L., 113; 20 Stat. L., 169, and 27 Stat. L., 593), the last being an act for the relief of entrants in good faith who had failed to meet the requirements of the original law. It was repealed in 1891 (26 Stat. L., 1095).

[11] Factors in the genesis of this memorial, and indeed public thought on forestry, not mentioned in preceding pages include Marsh's "The Earth as Modified by Human Action," the writings of Warder and Starr, and Brewer's work in connection with the Statistical Atlas of the United States, issued as a public document in 1874.

ruary 19, 1874.[12] Approval of the measure by the Secretary of the Interior and the Commissioner of the General Land Office was appended, and the resolution, with exhibits, was referred to the Committee on Public Lands in each house.

The Senate took no action, but the House committee issued a report[13] with a bill based upon the draft which had been submitted by the Association.[14] The new bill advanced no particularly new ideas. It called, merely, for the creation, in the Interior Department, of a Commissionership of Forestry, which office was to compile forestry statistics and make certain investigations. But the bill made no progress, nor did a similar bill introduced in the following Congress.[15] Eventually a rider was attached to the free-seed clause of the legislative, executive, and judicial appropriation act of August 15, 1876 (19 Stat. L., 143, 167), reading:

For purchase and distribution of new and valuable seeds and plants, sixty thousand dollars: *Provided,* That two thousand dollars of the above amount shall be expended by the Commissioner of Agriculture as compensation to some man of approved attainments, who is practically well acquainted with methods of statistical inquery, and who has evinced an intimate acquaintance with questions relating to the national wants in regard to timber to prosecute investigations and inqueries, with the view of ascertaining the annual amount of consumption, importation, and exportation of timber and other forest-products, the probable supply for future wants, the means best adapted to their preservation and renewal, the influence of forests upon climate, and the measures that have been successfully applied in foreign countries, or that may be deemed applicable in this country, for the preservation and restoration or planting of forests; and to report upon the same to the Commissioner of Agriculture to be by him in a separate report transmitted to Congress.

[12] 43 Cong. 1 sess., S. ex. doc. 28.
[13] 43 Cong. 1 sess., H. rep. 259. This included the memorial, Dr. Hough's paper, and other material on forest protection.
[14] 43 Cong. 1 sess., H. R. 2497.
[15] 44 Cong. 1 sess., H. R. 1310. Both bills were sponsored by M. H. Dunnell, of Minnesota.

The new organization, it will be noted, was placed under the Commissioner of Agriculture and had nothing to do with the timbered areas of the public domain, which remained under the General Land Office. In effect, a dual system had been set up, one department, Agriculture, being placed in charge of forestry without the forests, and the other, Interior, in charge of the forests without forestry, while the protection of the timber was still being carried on, at least theoretically, by the registers and receivers of the General Land Office under the circular of 1855.[16] This situation was destined to last for many years and result, in the end, in considerable controversy.

Research Initiated. Within two weeks after the approval of the act of August 15, 1876, Dr. Hough was appointed to conduct the forestry investigations.[17] The early appropriations were modest, the initial one of $2000 being followed by one of $2500 the next year. No appropriations were available for the years 1879 and 1880, however, and transfers of $2500 each year had to be made from other funds. In 1881 the forestry agency was reorganized into an administrative division, and from that time onward appropriations were continued each year without exception.

Dr. Hough, during the early years, gathered information through extensive correspondence, both in the United States and with forest officers abroad, the former dealing with such subjects as wood consumption in industry, and reasons for the unsatisfactory working of the Timber Culture Law. In 1878, officers at various army posts filled out questionnaires regarding local forest conditions.

The energies of the forestry service were devoted, for the most part during the early years under Dr. Hough, and, after

[16] Protection of live oak was practically abandoned in 1876.
[17] In 1875, at Chicago, the first forestry society in America was organized by Dr. John A. Warder—the American Forestry Association—though in 1874 an "American Forestry Council" had been formed in the American Institute of New York as a sort of standing committee on forestry for that organization.

1883, under Nathaniel Eggleston, to the interpretation of the data gathered. The results were published in the form of reports to Congress by the Department of Agriculture, which were combined in four volumes issued between 1877 and 1883.[18] In 1881 a special agent was sent abroad to observe forestry methods, and in 1883, 1884, and 1885 special agents were employed in investigating tree-planting progress in the prairie states, observing forest conditions in various other sections of the country, determining the causes of forest fires, etc.[19]

General interest in forestry was apparent during these years. The establishment of experiment stations in various forest regions to study conditions was urged; the American Forestry Congress[20] held a special meeting at the Department of Agriculture in May, 1884, at which numerous papers were read on various subjects, among others the influence of forest cover upon stream flow and climate; and the Forestry Division maintained an exhibit at the New Orleans exposition. In 1885 the Bureau of Entomology issued an extensive report on the ravages of forest insects,[21] and the Department of Agriculture sent out thousands of packets of seeds to prospective tree growers.

Much of the general interest displayed may be credited to the appearance in the mid-eighties of the "Report on the Forests of North America," by Professor Charles S. Sargent of Harvard.[22]

As an indication of the growing interest in and the importance of forestry the Senate Committee on Agriculture in 1884

[18] Dr. Hough prepared the first three and Mr. Eggleston the fourth.

[19] Commissioner Eggleston also urged that greater efforts be made to protect the timber lands on the public domain from depredations and, later, the withdrawal from sale and patenting of all such lands.

[20] Organized at Cincinnati in April, 1882. A "Southern Forestry Congress" was organized in Florida in 1885.

[21] Commissioner of Agriculture, Annual Report, 1885, pp. 319-32.

[22] 47 Cong. 2 sess., H. misc. doc. 42, part 9. This work was prepared for the Tenth Census, 1880, to which the author was attached as a special agent. It commanded wide attention and "gave rise to a very general discussion in the public press upon forests and their complex relations to the welfare and development of this country."—*Forest Leaves*, February, 1889, p. 23.

changed its name to the "Committee on Agriculture and Forestry."

A Change in Organization. On March 15, 1886, Dr. Bernhard E. Fernow succeeded Nathaniel Eggleston and on June 30, of the same year, the Division was given permanent statutory rank (24 Stat. L., 103). The new Chief of Division, in his first annual report, made this significant declaration:

It is not the control of the government over private property, it is not the exercise of eminent domain, it is not police regulations and restrictions that have produced desirable results upon private forestry abroad, but simply the example of a systematic and successful management of its own forests, and the opportunity offered by the government to the private forest owner of availing himself of the advice and guidance of well-qualified forestry officials.[23]

During the twelve years of Dr. Fernow's incumbency important forward steps were taken. In addition to the preparatory work in forestry history and statistics, which had occupied the first ten years under Hough and Eggleston, attention was given to subjects of more immediate application.[24]

The new lines of work which showed technical advancement (as distinguished from propaganda and compilations) included forest description, forest botany, forest influences, and, most important, forest products. Timber physics and properties and uses of wood were studied and reported upon, including data on railroad ties, heavy structural timbers, wood pulp, charcoal, and naval stores. Studies were made in tree-planting, forest mensuration, and even rain-making.[25] A complete forest

---

[23] Commissioner of Agriculture, Annual Report, 1886, p. 166.

[24] There was an American forestry exhibit at the Paris Exposition in 1900 (described in 51 Cong. 1 sess., H. ex. doc. 410, pp. 747-77) and at the World's Fair in Chicago (1893). Dr. Fernow had charge of both the American and German forestry exhibits at the latter exposition.

[25] The last was mandatory under the act of July 14, 1890 (26 Stat. L., 282, 286), and the Division proceeded to carry out the work, seemingly with little enthusiasm. The appropriation was continued the following year, but alarmingly heavy explosions used in experiments at Fort Myer, Virginia, across the river from Washington, appeared to terminate the

nomenclature, vernacular as well as botanical, was prepared by the Division dendrologist.[26]

Experiments in tree planting were hampered by conditions, however, the Division having no extensive forests to utilize for studies in silviculture and forest economy, nor, at the time, any corps of trained foresters. Private lands could not be used because the operating funds were public, and attempts to secure the use of military and other public reservations were unsuccessful. In 1891 certain tree-planting experiments were conducted in the sand hills of Nebraska, and in 1897 arid land tree-planting was tried at the request of the Secretary of the Interior.

Attempts were made for several years to further the work of the Division under the appropriation clauses which provided for the "distribution of valuable economic forest-tree seeds and plants," but such distributions proved to be futile and these attempts were abandoned.

In 1887 the first course of technical forest lectures for a body of students in America was conducted by Dr. Fernow at the Massachusetts Agricultural College. Before that time forestry lectures had been given as a part of the instruction offered by various land grant colleges. In 1898, the first forest school in America was established at Cornell University by the State of New York.[27]

---

program. It is interesting to note that as late as 1921 experiments in rain-making were revived, but they were individual and the Forest Service was not concerned.

[26] Mr. George B. Sudworth. This work analyzed over six thousand names and applied to five hundred American species. A monumental amplification of this study, " Check List of the Forest Trees of the United States," was published shortly before the death of Mr. Sudworth in May, 1927.

[27] Purchase of 30,000 acres for an experimental forest was also provided. It is to be recalled, however, that, in 1876, the executive committee of Amherst College urged the inclusion, in the curriculum, of courses in forest culture, while in 1880 the St. Paul Chamber of Commerce memorialized Congress for a grant of land for the establishment of a school of forestry. The latter movement was backed by Representative Andrews and Dr. Warder of the American Forestry Association. It was opposed by President Eliot and Professor Sargent of Harvard, who maintained that the time was not ripe for such a venture. The project failed.

On the side of developing a technique in an American practice of forest management, less may be said for the twelve years
under discussion. The first bulletin, the title of which suggests
practical information relating to the actual handling of forests
or growing of trees, was " Some Foreign Trees for the Southern
States (Cork, Wattle Tree, Eucalyptus, Bamboo)," which did
not appear until 1895,[28] and even this was somewhat remote
from establishing the practice mentioned. Dr. Fernow himself
characterized the published material up to the close of his
incumbency as one-half original and one-half compilation.
However, looking forward, also, he stated:

I may be permitted upon my retirement . . . to characterize
the past period of twenty years . . . as *the period of propaganda and primary education* . . . we have laid the foundations. . . . While at first the Division of Forestry was the only
educational element in the forestry movement, it may now, perhaps, be left to other agencies to carry on a general campaign
and propaganda of enlightenment and *the Division can concentrate itself more upon the development of the technical side
of forestry.*[29]

The growth of the Division during the Fernow administration was not rapid, though encouraging as compared with the
previous ten years. From 1876 to 1898 the personnel increased
from one employee to eleven (including two trained foresters).
Appropriations from 1876 to 1886 totalled $60,000; from 1886
to 1898, $230,000. But public interest had not abated and
legislation and attempted legislation were active.

Legislation. Between 1868 and 1871 some dozen tree planting bills were introduced in Congress, with three more in 1872.[30]

[28] See Fernow, Report upon forestry investigations, pp. 40-44. This lists all
publications from 1877 to 1898.

[29] *Ibid.*, pp. 26, 27. Italics not in original.

[30] Two of the latter required homesteaders on forested tracts to retain
a certain proportion in timber. See 42 Cong. 2 sess., H. R. 2197; 42 Cong.
3 sess., H. R. 3008; and 42 Cong. 2 sess., S. 680. This last, introduced by
Senator Hitchcock of Nebraska, February 20, 1872, became, in the following session of the same Congress, the Timber Culture Law.

Between 1870 and 1898 more than 150 forestry measures were introduced in Congress.

Certain of these laws and bills deserve comment. The naval appropriation act of March 3, 1871 (16 Stat. L., 526, 527), provided $5000 "for protection of timber lands." Primarily this was intended for the protection of naval timber reservations, and it was principally expended for that purpose. Nevertheless, this constituted the first appropriation made directly for the protection from spoliation of publicly owned timber in the United States, and the wording was broad enough to permit a more comprehensive guardianship than was exercised. The agents appointed were expected to have an eye, also, for the protection of adjoining unreserved public lands,[31] but their prime duties were with the naval reservations.

The evidence shows that the above mentioned law[32] did not contemplate a general scheme of timber protection throughout the country, and hence does not deserve the distinction, sometimes accorded it, of being the first appropriation for the general protection of all publicly owned timber. Such distinction fell rather to a law passed a little over a year later. This was the sundry civil appropriation act of June 10, 1872 (17 Stat. L., 347, 359), which made available $10,000 for the protection of public lands in general.[33]

The Timber Culture Act of March 3, 1873 (17 Stat. L., 605), was enacted to promote tree planting in the treeless regions, but it attained, in practice, almost negligible results, though lending itself admirably to the abuse of appropriating public land for private uses. Of similar undesirable outcome

[31] The naval timber agents had probably been expected to do this even before 1871; that is, after the Supreme Court, in 1850, had interpreted the act of 1831 as applying to *all* timber lands. United States v. Briggs, 5 Howard, 208, and 9 Howard, 356.

[32] See "Statement of appropriations and expenditures of the Navy Department from March 4, 1789, to June 30, 1876," p. 68; 45 Cong. 1 sess., S. ex. doc. 3; Secretary of the Navy, Annual Report, 1872, pp. 71, 72; General Land Office, Annual Report, 1877, p. 20.

[33] There was an unsuccessful attempt at an amendment providing for two special agents to aid in "preventing depredations" and "prosecuting trespassers."

were the Relinquishment Act of June 22, 1874 (18 Stat. L., 194), and the several acts culminating in the general Right-of-Way Act of March 3, 1875 (18 Stat. L., 482).[34]

The year 1876, as has been noted, saw the creation of the federal forestry agency, but this year also witnessed other significant activities. Colorado became a state, and its constitutional convention memorialized Congress, asking for the transfer of the public timber lands in the territory to the care and custody of the state. The new constitution also provided that the general assembly should enact laws to preserve the forests on state lands or upon public domain lands placed under its control.

A bill containing interesting features was introduced in Congress, also, in 1876. It was " for the preservation of the forests of the national domain adjacent to the sources of the navigable waters and other streams of the United States," [35] which, if the naval live-oak acts be ignored, constituted the first attempt to create national forests in America. This bill proposed setting aside and preserving all .public lands " adjacent to sources and affluents of all rivers " in order to prevent such rivers becoming " scant of water," and is of added interest because of its similarity, in principle, to the Weeks Act of 1911. The bill failed of passage a few months before the creation of the new forestry agency.

In 1877 the State of Connecticut sent a commissioner to Europe to study forestry methods, and in January 1878 a bill was introduced in Congress which went still further than the unsuccessful one of 1876, above mentioned. This bill [36] provided for the withdrawal of all timbered lands from settlement and for their administration and protection by a forester and a corps of assistants, who were empowered to sell timber as needed but under careful cutting restrictions that would preclude destruc-

[34] General Land Office, Annual Report, 1885, pp. 28, 29, 83, 311, *et seq.;* Secretary of the Interior, Annual Report, 1885, p. 41; 1886, p. 29.
[35] 44 Cong. 1 sess., H. R. 2075.
[36] 45 Cong. 2 sess., S. 609.

tive denudation. It also provided for the sale of agricultural lands in the reserved timber areas at a price to cover timber as well as land, prescribed penalties for setting fires, and provided that suits for trespasses committed on the public lands (save naval reservations) should be brought by the Secretary of the Interior.[37] This bill likewise failed to become a law.

On June 3, 1878, came two more acts which were passed over the vigorous opposition of the Secretary of the Interior. These were the " Free Timber " and " Timber and Stone " Acts (20 Stat. L., 88 and 89). The first of these laws, among other things, repealed Section 4751 of the Revised Statutes (at least so far as Colorado, Nevada, and the territories were concerned),[38] an end sought in the bill of the preceding January. In other respects, however, this new legislation failed to carry on the progressive ideas of the unsuccessful bills mentioned. It contributed, rather, to unrestricted exploitation, being sponsored by influential interests which profited by existing conditions, and backed by political power representing the lax sentiment prevailing in the regions affected.[39]

The entire question of forest protection was brought up anew by the sundry civil appropriation act of March 3, 1879 (20 Stat. L., 377, 394), which provided for a study of the entire public land question by a special commission.[40] The report of this Commission,[41] which included a proposed law covering all

[37] This last, amounting to an amendment of R. S., 4751, had been urged by Secretary Schurz in his 1877 report (p. 18). R. S., 4751 was Section 3 of the act of 1831, which provided for bringing all timber trespass suits under direction of the Secretary of the Navy. This brought about an anomalous situation, since most of the timber lands were under jurisdiction of the Secretary of the Interior. Section 4751 was substantially amended along the lines attempted in the bill of January 28, above mentioned, by the acts of April 30, 1878 (20 Stat. L., 46), and June 3, 1878 (20 Stat. L., 88).

[38] It was repealed for all " public-land states " by the act of August 4, 1892 (27 Stat. L., 348).

[39] The first of these laws permitted the felling and removing of timber on the public domain (in Colorado, Nevada, and the territories) for mining and domestic purposes, and the second, the sale of timber and stone lands in California, Oregon, Nevada, and Washington.

[40] Composed of the Commissioner of the General Land Office, the Director of the Geological Survey, and three private citizens appointed by the President.

[41] 46 Cong. 2 sess., H. ex. doc. 46.

phases of public land administration and disposal, recommended the withdrawal of all timber lands from sale or other
disposal, the sale of public land timber for commercial purposes and its free use under certain conditions, and the administration of the public timber lands by the Commissioner of the
General Land Office. Penalties were proposed for timber spoliation which were similar to those existing in the old live-oak
statutes.[42] Legal action was to be initiated by the Secretary of
the Interior, who was given authority to " compromise," and it
was proposed to " condone and quash " all civil and criminal
actions then pending.

With one exception the recommendations of this Commission
failed to become law, and that exception was the condoning and
quashing of pending actions, which was provided for in the act
of June 15, 1880 (21 Stat. L., 237). By this act all trespasses
committed prior to March 1, 1879, were to be " compromised "
upon payment of the government price on the land involved,
namely, $1.25 per acre. After that date all trespass acts were
to be rigorously enforced.

State interest in forestry continued unabated. The effective
beginnings of the Adirondack and Catskill forest preserves
in New York date from 1885. In 1885, also, California created
a state board of forestry, which urged in its first annual report
that all government and state timberlands not fit for agriculture be permanently reserved, and that the cutting of timber
be placed in the hands of national or state forestry officers. In
1888 the California legislature passed a resolution praying
Congress to stop the disposal of all government lands in California with a view to their permanent preservation as a forest
reserve for the protection of the watersheds of the state.

In January, 1888, Senator Hale, of Maine, introduced in
Congress, at the request of the American Forest Congress,
a bill[43] which provided for the establishment and management

---

[42] R. S., Sections 2460-63 and 4751. These provisions, the Commission
specifically declared, should not be repealed by the proposed act. The new
law was to have " general applications " as contrasted with the old.

[43] 50 Cong. 1 sess., S. 1476 and S. 1779.

of forest reservations under a commissioner of forests in the, Department of the Interior. This bill aroused considerable opposition, one senator protesting against referring the bill to the Committee on Agriculture and Forestry and declaring that forest protection, to be effectual, would always have to be under the direction of the Secretary of the 'Interior, through the medium of the Commissioner of the General Land Office.

In 1888, also, the agitation which had been aroused by the swindling of the Indians with regard to timber on their reservations reached a climax. During the sixties and seventies this practice had been prevalent, and finally one culprit was tried under Section 5388, R. S. (Act of March 3, 1859; 11 Stat. L., 408). The defendant maintained, and was upheld by the court, that this law did not apply to Indian lands, and that, hence, no conviction was possible. In passing on the case the court took occasion to remark, however, that ". . . the duty of Congress . . . to protect these Indians from unlawful intrusion from without and from violations of their rights by any and all persons is manifest." [44]

But though Congress did not act for some years, agitation was not lacking. Four special messages by Presidents Arthur and Cleveland, between 1882 and 1888, urged enlargement of the law,[45] as did a report in this period.[46] Finally, the complaints of depredations resulted in a Senate committee investigation. This began in March, 1888, and terminated with a voluminous report,[47] which was supplemented, December 22, 1890, by a special message from President Harrison.[48] Numerous pieces of legislation resulted from this investigation, the act of January 14, 1889 (25 Stat. L., 642), concerning the Minnesota Chippewas, being, perhaps, typical. This provided for the

[44] United States v. Reese, 5 Dillon, 405; 46 Cong. 2 sess., S. misc. doc. 100.
[45] 47 Cong. 1 sess., S. ex. doc. 89; 48 Cong. 1 sess., H. ex. doc. 14; 49 Cong. 1 sess., S. ex. doc. 13; 50 Cong. 1 sess., S. ex. doc. 42
[46] 47 Cong. 1 sess., S. rep. 392.
[47] 50 Cong. 2 sess., S. rep. 2710.
[48] 51 Cong. 2 sess., S. ex. doc. 23.

cession of all Indian lands to the government except enough to provide each Indian with an allotment, and for the purchase of all pine thereon at a full and fair price under strict government supervision, after careful survey and appraisal. Resulting moneys were to go into a fund for the benefit of the tribe.

Meanwhile, in every Congress succeeding the report of the Commission of 1879, consideration was given to measures dealing with the revision of land laws along lines at least approximating those recommended by the Commission. On January 20, 1890, President Harrison transmitted to Congress, with a special message, another memorial from the American Association for the Advancement of Science, urging a thorough study of the forest situation by an expert commission and the withdrawal of all forest lands from sale and entry pending its completion.[49]

The varying tides of bill and law and the cumulative effect of public opinion now converged, and the result was the vital and fundamental act of March 3, 1891 (26 Stat. L., 1095, 1103), which repealed the Timber Culture Law and the Pre-emption Law, amended the Homestead and Desert Land laws with a view to making them less susceptible to fraud and manipulation,[50] and abolished public sales of government lands.[51] An amendment passed the same day (26 Stat. L., 1093) established the permit cutting system.

The most important provision, however, from the standpoint of the development of government forestry in America, was contained in Section 24 of the act, which had been inserted only at the eleventh hour. This section provided:

That the President of the United States may, from time to time, set apart and reserve, in any State or Territory having public land bearing forests, *in* [*sic*] any part of the public lands

[49] 57 Cong. 1 sess., S. ex. doc. 36.

[50] By making commutation possible only after fourteen months' residence and cultivation.

[51] Cash sales had been done away with, partially, by the act of March 2, 1889 (25 Stat. L., 854).

wholly or in part covered with timber or undergrowth, whether of commercial value or not, as public reservations, and the President shall, by public proclamation, declare the establishment of such reservations and the limits thereof.

Here was the foundation upon which were to be built up the first true national forests in America,[52] and the act of which it formed a part was one of two laws[53] without which no national forests in America would have been possible.

Within a few months of the enactment of this law the President had withdrawn some 2,500,000 acres of timber land in Wyoming and Colorado, and within the next two years the withdrawals increased to a total of over 17,500,000 acres, not including the national parks.

The new law was a step in the right direction, but it was only a step. While the forests could no longer pass into hands of speculators, as had been possible under the Timber and Stone and other acts of similar nature, yet the forest reserves were still without any provision for protection and administration and hence were practically unprotected against ordinary thievery, fire, and unrestricted grazing. The unreserved forests, under the permit cutting clause, previously mentioned, were opened even more widely to speculative activities. The pertinent provisions follow (Section 8 as amended):

. . . in the States of Colorado, Montana, Idaho, North Dakota and South Dakota, Wyoming, and the District of Alaska, and the gold and silver regions of Nevada and the Territory of Utah in any . . . action . . . for a trespass . . . or to recover timber or lumber cut thereon it shall be a defense if the defendant

[52] The reserves set up under this law could not be called, as yet, true national forests, however. Previous reservations of the public domain for purposes bearing some resemblance to forest conservation had been live-oak reservations for naval purposes, Yellowstone, and one or two other of the earlier national parks, and the various wood and timber reservations set aside from wooded areas adjacent to certain army posts to supply fuel to such posts. None of them was a reservation looking to preservation of the general national timber supply, through either ordered use or timber hoarding; hence, each had little, if anything, in common with national forests as that term is properly understood.

[53] The second was the act of 1897.

shall show that the said timber was so cut or removed from the timber lands for use in such State or Territory by a resident thereof for agricultural, mining, manufacturing, or domestic purposes under rules and regulations made and prescribed by the Secretary of the Interior and has not been transported out of the same, . . . the Secretary of the Interior may make suitable rules and regulations to carry out the provisions of this act, and he may designate the sections or tracts of land where timber may be cut, and it shall not be lawful to cut or remove any timber except as may be prescribed by such rules and regulations, but this act shall not operate to repeal the act of June third, eighteen hundred and seventy-eight, providing for the cutting of timber on mineral lands.

Attempts to remedy the deficiencies of the law continued during the six years from 1891 to 1897. Two bills introduced during this period by Senator Paddock, of Nebraska, and Representative McRae, of Arkansas, respectively, deserve comment.[54] Both were comprehensive forestry measures and both were strongly recommended by the Secretary of the Interior. The McRae Bill provided for protective administration by the Secretary of the Interior, regulated the sale of lumber *apart* from the land, and restored to entry, lands primarily valuable for agriculture. It also provided for the use of the army to prevent thievery, harmful grazing, etc. Neither bill passed and the American Forestry Association therefore asked the Secretary of the Interior to request the National Academy of Sciences to investigate the entire forestry question and report " upon the inauguration of a rational forest policy for the forested lands of the United States." Suggestions were also invited for proper legislation to remedy specific, existing evils. The Academy appointed a commission in response to this request, which began work in July, 1896, the report being submitted in May, 1897.[55] The recommendations included, substantially, the provisions of the unsuccessful McRae Bill. In the meantime, however, the

[54] 52 Cong. 1 sess., S. 3235 and 53 Cong. 1 sess., H. R. 119. The Paddock Bill was advocated by the American Forestry Association.
[55] 55 Cong. 1 sess., S. doc. 105.

commission had urged the creation, forthwith, of thirteen additional forest reserves, including an area of more than 21,000,-000 acres, and President Cleveland, on February 22, 1897, issued a proclamation setting aside the areas designated. This action was so precipitate that it aroused the most vigorous and widespread opposition in the West. Seemingly the President did not consult, or even notify, the representatives of the states affected. Even the forestry leaders of the country deplored the suddenness of the action because of the unfortunate psychological possibilities it held as an argument against the reservation policy. In spite of sectional objections, however, Congress passed the vitally important act of June 4, 1897 (30 Stat. L., 11, 34), which included the main features of the McRae Bill, in addition to many other important provisions for which the American Forestry Association had long striven. Legislative foundation was here laid for the establishment, in the Department of the Interior, of a protective and administrative organization for the forest reserves permitted by Section 24 of the act of 1891 and set aside, from time to time, by the President.

The new law provided for the regulated utilization of the reserved forest areas as well as for their protection against fire and trespass. Affirming all previous executive orders and proclamations setting aside wooded areas of the public domain as forest reserves, it stipulated that no such reservations should be made except for certain definite purposes: the improvement and protection of the reserved forests, the "securing [of] favorable conditions of water flow," and the provision of "a continuous supply of timber for the use and necessities of the citizens of the United States." These were basic and vital provisions.

The act also included some undesirable features. The first was the stipulation that the extensive reservations made by President Cleveland, in the preceding February, might not take effect for nine months after the passage of the law. California was not included. A little less than 20,000 acres passed into

3

private hands as a result of this suspension. The second was the so-called "forest-lieu" clause, which was intended to relieve settlers who found their homesteads surrounded by a forest reservation and enabled them to trade their land for a tract in the unreserved public domain. This resulted in a widespread trading of worthless railroad grants and denuded timber lands for valuable unreserved tracts on the public domain. The third undesirable feature of the law was the "non-export" clause, which forbade the shipping of lumber from the reserves out of the state of origin. Except for these provisions, the new law was admirable and led to admirable achievement.

The Forests. On January 19, 1854, " all the letters and other papers that had hitherto been filed in the Department [of the Interior] in relation to depredations committed on the public lands " were turned over to the General Land Office.[56]

On January 28, 1854, the Commissioner of the General Land Office issued a circular to timber agents in the field which informed them that the main object of their appointment was "the prevention of waste or trespass on the public lands and the destruction or carrying away of the public timber," and warned them against harassing *bona fide* settlers, compounding or compromising with trespassers, and pretending to consummate *bona fide* settlements as a cloak to spoliation. These warnings were the result of frequent accusations that such things had been done under the system inaugurated in 1850. On March 4, 1854, timber agents were put under $20,000 bond. Meanwhile, under Commissioner John Wilson, enforcement had been rigorous in the extreme and the opposition was likewise so vigorous that means were finally found to remove the enforcer and appoint a new Commissioner of the General Land Office, on August 3, 1855. The new officer sent out a circular on December 24, 1855, containing the statement that " The Secretary of the Interior has concluded to change the present system

[56] General Land Office, Annual Report, 1877, p. 16.

of timber agencies, and to devolve the duties connected therewith upon the officers of the local land districts."

These officers had other important duties and they were to be paid no more for their additional responsibilities. Hence they were instructed, when trouble arose, to " refer the facts to this office [General Land Office, at 'Washington] for consideration and await instruction." They might, in " pressing emergency" delegate their authority to deputies, but they were required to " report the fact instanter, and the necessity for it." The result was a general centralization of authority and action at Washington and a firm stand on the practice of " compromising " with timber offenders.

This stand was not firmly maintained, however, and violations led to practises finally deemed wise and incorporated in the law. As early as 1860 compromises had been entered into, seemingly with the full approbation of the Secretary of the Interior,[57] in spite of the amplification of the trespass section of the act of 1831 by the act of March 3, 1859 (11 Stat. L., 408).[58] Two years later the same officer took issue with the no-compromise policy, refusing "to concur . . . in the opinion that no settlement is to be made with trespassers" and indicating that if the government could get the price of the land out of the stolen timber the rest did not matter.[59] The report of the General Land Office in 1864[60] deprecated the pursuit of offenders " in a vindictive spirit," and asserted that the exaction of a "liberal stumpage" (or a " reasonable tariff " or "revenue") had resulted in mitigating the trespass evil and returned to the treasury " over ten thousand dollars." Nevertheless, the 1865

[57] See General Land Office, Annual Report, 1860, p. 18.

[58] The act of 1831, it may be noted, covered naval reservations and the general public domain. The act of 1859 extended protection to lands " reserved or purchased for military and *other* purposes."

[59] General Land Office, Annual Report, 1862, p. 20; 1877, p. 18. See also Wells v. Nickles, 104 U. S. 444.

[60] P. 21. There had been definite question raised as to the legality of enforcement in the middle fifties and the Commissioner in 1865 believed the duty to be " legitimately incidental."—Annual Report, 1865, pp. 26, 27.

report definitely stated: " The timber belongs to the United States and no authority to sell or to permit anyone to cut or use it exists."

The 1866 report declared that timber protection was an " incident to the land administration," and the 1872 and 1873 reports defended " compromise " and " stumpage " as a source of revenue and a means of avoiding " vindictiveness " toward citizens. By 1876, then, compromise, on the basis of " a reasonable stumpage according to the market value of the timber cut," had become a regular and customary procedure. Local officers appear to have been engaging in it without question, and, in many cases, the taking of timber seems to have been arranged openly in previous agreements between the lumbermen and General Land Office officers.

Reaction against this practice took form in 1872 with the attempt to append an amendment to the sundry civil bill for 1873 providing for two special agents to act under the Commissioner of the General Land Office for the purpose of assisting " registers and receivers in preventing depredations upon the timber laws of the United States, and in prosecuting trespassers thereon." [61]

The new Commissioner urged the complete removal of timber lands from the operation of the pre-emption and homestead laws, and the wholesale disposal of such lands, by sale, after careful survey and appraisal.

The bill mentioned above (without the amendment) became law on June 10, 1872 (17 Stat. L., 347, 359), and carried $10,000 as a direct appropriation for the protection of the timber lands,[62] but " compromise " continued as a policy until 1877, when Carl Schurz became Secretary of the Interior. Just previous to the accession of Schurz, Commissioner Williamson of the Land Office had directed the registers and receivers to

[61] Cong. Globe, June 7, 1872, pp. 4360, 4361; 42 Cong. 2 sess. This assistance had been requested by the Secretary of the Interior because of reports of widespread frauds in the timber lands.

[62] Succeeding years continued this practice without exception; 1873, $8000; 1874, $5000; 1875, $5000; 1876, $5000. Fifty years later (1926) the amount was $430,000.

employ no more timber deputies without previous authorization from him and required them to submit a statement of the
timber business phase of their respective offices from 1855 on.
The facts thereby obtained, and a general study of the situation, resulted in the recommendation that registers and receivers
should be relieved of the timber responsibilities and for that
work a force of special agents should be appointed by the Commissioner and paid out of the appropriations instituted in 1872.
Possession of authority to do this was based upon two provisions of law and a court decision.[63] The Secretary of the Interior
approved the plan, and on May 2, 1877, it was put into effect.[64]
Timber protective work was to be performed, thenceforward,
by employees specially delegated from Washington who were
forbidden to "compound or compromise" with trespassers.
Local officers were directed to coöperate with timber agents
upon request, the latter being authorized and instructed to con-
fer with, and seek the aid of, federal district attorneys when
need arose.

The small available appropriation was supplemented by con-
tingent funds of the General Land Office, an organization for
the suppressing force was created, and a policy of vigorous
enforcement was initiated. The drive was uncompromising.
Offenders faced criminal as well as civil action.[65] Persons who
desired to cut timber on the public lands for use in mining or
other legitimate purposes were warned of prosecution.[66] This
vigorous policy did not continue, however, under the succeeding administration.

[63] R. S., 453 directing the Commissioner to perform all executive duties
respecting the public lands (Sec. 699, U. S. Comp. Stats., 1918); the current
appropriation clause (19 Stat. L., 122) "To meet expenses of suppressing
depredations upon the public lands . . ."; and decision U. S. Supreme
Court, United States v. Cook (19 Wallace, 591) holding that the timber,
while standing, is a part of the realty, and it can only be sold as the land
could be, and unless lawfully cut will remain the property of the United
States."

[64] The circular of 1855 was revoked. For full text of the 1877 circular, see
General Land Office, Annual Report, 1877, p. 21.

[65] Secretary of the Interior, Annual Report, 1877, p. 17.

[66] 45 Cong. 2 sess., S. ex. doc. 9, pp. 133, 147. Congress had not taken
action to make cutting of any sort legal until 1878.

While the acts of 1878 and that of 1880, previously discussed, affected certain phases of trespass prosecution, made certain cutting possible, and "compromised," legally, certain offenses, the timber protection administration under the General Land Office continued as in the past until 1898. The act of 1891 while permitting the creation of forest reserves had not changed the method of administration or protection. With the act of June 4, 1897 (30 Stat. L., 34), came organization changes. On June 30, 1897, the Commissioner of the General Land Office, acting for the Secretary of the Interior, had issued rules and regulations for the government of the forest reserves set apart under the act of 1891, but the organization to carry out such rules and regulations had to wait upon an appropriation of $75,000, which did not become available until July 1, 1898 (30 Stat. L., 618). This was $35,000 less than the appropriation for the maintenance of the special agent system in the same year.

The new machinery was planned along the lines of the existing force of special agents for the protection of public domain timber from theft, that is, on the district basis. The reserves were divided among eleven districts, each district under a superintendent, while the districts, in turn, were divided into reservations in charge of supervisors. Each supervisor had a force of forest rangers to perform detailed work.

For some time the Commissioner directed this work through the instrumentality of the Special Service Division in the General Land Office, but in 1901 the Forestry Division of the General Land Office was created, and a trained forester was placed in charge,[67] with three other foresters to assist him. As thus constituted the Division continued until the major changes of 1905 came about.

---

[67] Trouble (trespass) had developed in the Black Hills reserves, and upon request of the Secretary of the Interior a man was assigned by the Bureau of Forestry of the Department of *Agriculture* to clear up the situation. This agent reported to the Secretary of the Interior, however. Between October, 1901, and February, 1902, this agent (the second; the first had resigned) framed the first timber sale contract ever made in the national forests.

Developments, 1898-1905. When Gifford Pinchot became head of the Division of Forestry of the Department of Agriculture in 1898, a new era had begun, though consummation of the desired organization was to wait for seven years. In 1898 forest reserve administration was still in the Department of the Interior and the Division of Forestry of the Department of Agriculture had no jurisdiction over such work. A program to meet this anomalous situation was announced as follows:

To introduce, in practice, better methods of handling forest lands of private owners, including both wood lots and large areas chiefly held for lumber, and afterwards to spread a knowledge of what had been accomplished.

To assist the Western farmer to plant better trees in better ways.

To reduce the loss from forest fires.

To inform . . . citizens regarding opportunities for forest enterprises in Alaska, Cuba, and Porto Rico.

This new policy was shaped both to draw private lumbering into conservation practices and to train the personnel in practical forest administration.

Offers of coöperation, expert advice, and assistance were made to the owners of private timber tracts and the response was immediate, and wide-spread. "Working plans" were prepared for private interests and owners and by 1905 there were 900,000 acres of private timber lands in seven states being managed along approved forestry lines, and applications for advice from owners contemplating forestry management covered over 2,000,000 acres more. More significant, as a result of the new plans, was the statement of the Secretary of Agriculture in 1904 that the sentiment of lumbermen had changed from suspicion to support; that of the general public from indifference to interest.

Certain public work was also undertaken. The Division, as official forestry adviser in technical matters to the Department of the Interior, drew up working plans for several of the national forests, beginning with the Black Hills reserve in 1900.

Similar relationships were established with the War Department in connection with its wood and timber reservations, and with the Philippine Forest Bureau.

At the beginning of the fiscal year 1900 (Act of March 1, 1899, 30 Stat. L., 949), the designation of the Chief of the Division of Forestry was changed to Forester. Two years later (Act of March 2, 1901, 31 Stat. L., 929) the Division became the Bureau of Forestry.

The laws passed regarding the use of timber on Indian reservations had furnished the opportunity for the first important practical development of American forestry and the year 1902 saw the initiation of these plans. A law approved June 27, 1902 (32 Stat. L., 400), further amended the Indian forest act of 1889, and while it did not alter the underlying purpose of the latter, it strengthened it by providing for the utilization of forest knowledge in working out its provisions.

As to such lands, in general, it permitted the removal of only merchantable pine, prohibited the cutting of more timber than was absolutely necessary in the economical conduct of logging operations, and required the burning or removal of enough of the debris remaining after logging to minimize the fire hazard. For some 200,000 acres the law went further. It provided that logging should be carried on solely in accordance with plans and specifications prescribed by the Forester (Agriculture) with the approval of the Secretary of the Interior and that five per cent of standing pine should be left for reforestation purposes. Such special areas were to be selected by the Forester and exempted trees were to be designated under his direction. The act provided, further, that, after logging, the special area was to become a forest reserve as if created in accordance with the act of 1891. The Bureau of Forestry was thus given its first major experience in practical forest work on government land.

Finally this act of 1902 made provision for setting aside from the Chippewa lands in Minnesota " ten sections, islands

and points," which were henceforth closed to settlement and
to sale of land or timber. Such sections were to be selected by
the Forester with the approval of the Secretary of the Interior.
(Some dispute over jurisdiction having arisen in this connection between the Bureau of Forestry and the Department of the
Interior, the latter asserted its jurisdiction over the sections in
question and over forest reservations in general.) The Bureau
also made studies of the Chippewa reservations in Wisconsin
and, at the request of the Secretary of the Interior, drew up
plans for scientific lumbering and for fire protection.

The work of the Bureau had now become well organized,
comprising activities in: (1) Forest management, (2) forest
investigations, (3) tree-planting, and (4) records.

The work in forest management has been discussed. The
forest investigations involved work of a research nature and
attempts at solution of various forest problems. They were
conducted both independently and in coöperation with other
bureaus. An example of results obtained was the development
of the "cup and gutter" system for extracting turpentine,
which did away with the injurious and wasteful "boxing"
system previously in use.

Studies were also made of the use of tree planting to check
sand-dune encroachment; timber testing work, previously sus-
pended, was resumed in coöperation with the Bureau of Chemistry; and paper pulp investigations were made with the same
bureau. Forest insect studies were initiated with the Bureau
of Entomology.

In 1901, in coöperation with the Geological Survey, the
examination of 9,600,000 acres of forest land in the Southern
Appalachian region was begun, to ascertain its suitability for
a forest reserve. This was the first definite step toward the
adoption of a policy of forest protection for eastern watersheds which was finally enunciated by the Weeks Act of 1911.
Agitation for this policy had begun in 1899 with the organization of the Appalachian National Park Association in North

Carolina. An exhaustive report of this coöperative investigation was made by the Secretary of Agriculture in December, 1901.[68]

One or two pieces of contemporary legislation also deserve comment. The acts of June 6, 1900 (31 Stat. L., 588, 614), and March 3, 1901 (31 Stat. L., 1037), restricted selections made under the "forest lieu" clause of the 1897 law, to vacant, non-mineral, surveyed public lands which were subject to homestead entry. They gave lieu selectors an extra period of grace, however, until October 1, 1900, before which they might select from unsurveyed as well as surveyed lands. This resulted in land frauds in Oregon and California. The Secretary of the Interior had initiated investigations in 1901, but the information became public in 1902, and focussed attention on the timber lands.

The act of March 3, 1905 (33 Stat. L., 1264), abolished the "forest lieu" privilege as a matter of general application. Meanwhile, in 1903, President Roosevelt had appointed his Public Lands Commission to examine into the public land laws. The Forester sat on this Commission, which, after careful inquiries, issued two partial reports in 1904 and 1905.[69]

While no laws resulted directly from the recommendations in this report it unquestionably blazed the trail for noteworthy conservation advancement a few years later, and is generally regarded as the real genesis of the conservation movement. Meanwhile, the development of the work of the Bureau had been rapid as had the increase in number and area of the forest reserves. The Bureau also was assuming an increasingly important rôle as initiator of proclamations for the establishment of new national forests.

The growth of the Bureau organization was noteworthy. From an enrollment of eleven employees in 1898 it had grown to 821 by 1905. The increase, in appropriations, is also informing: $28,520 in 1898 to $439,873 in 1905. The total for

[68] 57 Cong. 1 sess., S. doc. 84.
[69] 58 Cong. 3 sess., S. doc. 189.

the period 1898 to 1905 was $1,461,253, compared with $290,-000 for the period 1876 to 1898.

The anomaly of organization concerned with forest affairs persisted, the work being divided among three bureaus in two departments. To this situation President Roosevelt now turned his attention and in his annual message of December 9, 1904, he recommended centralization under the Bureau of Forestry in the Department of Agriculture. Within six days a bill[70] embodying such changes as were necessary passed the House and a month later was reported in the Senate with a strengthening amendment.

Meanwhile, the American Forest Congress, called by the American Forestry Association, convened in Washington to promote the establishment of " a broader understanding of the forest in its relation to the great industries depending upon it; to advance the conservative use of forest resources for both the present and future needs of these industries; to stimulate and unite all efforts to perpetuate the forest as a permanent resource of the nation." The Secretary of Agriculture acted as presiding officer, and the President of the United States in his address declared that the object of forestry was not to "lock up" forests but to " consider how best to combine use with preservation."

The Congress was attended not only by the leading exponents of forestry, but also by leading lumbermen and railroad men and large-scale users of forest products. It adopted resolutions urging: the repeal of the Timber and Stone Act; amendment of the Forest Lieu Act; the creation of forest reserves in the eastern watersheds; and unification of all government forest work in the Bureau of Forestry of the Department of Agriculture.[71]

The effect of this convention upon pending legislation is generally recognized. The bill previously referred to became law on February 1, 1905 (33 Stat. L., 628), and took effect

[70] 58 Cong. 2 sess., H. R. 8460.
[71] American Forest Congress, Proceedings, 1905.

immediately. During the same session the designation of the Bureau of Forestry was changed to the Forest Service, effective July 1, 1905 (Act of March 3, 1905; 33 Stat. L., 861, 872).

**The Rise and Development of " United States Forestry,"
1905-1920.** Prior to the passage of the act of 1905 the forest work of the government was allocated among at least four separate units. The Bureau of Forestry of the Department of Agriculture, the Forestry Division of the General Land Office of the Department of the Interior, and the corps of special protective agents, also of the Land Office, have been discussed. The fourth unit was the Geological Survey of the Department of the Interior, which under the act of 1897 was charged with the survey and mapping of the forest reserves.[72]

This work was not transferred to the new organization, but with this exception, one forestry unit in a single department was now to perform all of the forest work of the national government everywhere.

The new law directed immediate transfer of the forest reserves—a total area of over 63,000,000 acres—to the jurisdiction of the Department of Agriculture and provided that revenues derived from the reserves during a period of five years from the passage of the act should be expended for the protection, administration, improvement, and extension of such reserves as the Secretary of Agriculture might direct.

The administrative organization existing in the Division of Forestry of the General Land Office, comprising some five hundred employees, was merged into the older unit without radical change.

---

[72] The Geological Survey, from its beginning in 1879, had gathered forest data and was able in 1891 to give considerable assistance as to locating boundaries in setting aside the new reserves. Later regulations (1897-1905) issued by the General Land Office for administering reserves made extensive use of information secured by the Division of Geography and Forestry. This Division gathered data concerning some seventy million acres of forest and collected information regarding general forest affairs apart from the reserves.

On the day the law went into effect the Secretary of Agriculture issued the following instructions to the Forester regarding the general principles to be followed in administering the forests under the new régime:

In the administration of the forest reserves it must be clearly borne in mind that all land is to be devoted to its most productive use for the permanent good of the whole people and not for the temporary benefit of individuals or companies. All the resources of forest reserves are for use, and this use must be brought about in a thoroughly prompt and business-like manner, under such restrictions only as will insure the permanence of these resources. The vital importance of forest reserves to the great industries of the Western States will be largely increased in the near future by the continued steady advance in settlement and development. The permanence of the resources of the reserves is therefore indispensable to continued prosperity, and the policy of this Department for their protection and use will invariably be guided by this fact, always bearing in mind that the *conservative use* of these resources in no way conflicts with their permanent value. You will see to it that the water, wood, and forage of the reserves are conserved and wisely used for the benefit of the homebuilder first of all; upon whom depends the best permanent use of lands and resources alike. The continued prosperity of the agricultural, lumbering, mining and live-stock interests is directly dependent upon a permanent and accessible supply of water, wood, and forage, as well as upon the present and future use of these resources under business-like regulations, enforced with promptness, effectiveness, and common-sense. In the management of each reserve local questions will be decided upon local grounds; the dominant industry will be considered first, but with as little restriction to minor industries as may be possible; sudden changes in industrial conditions will be avoided by gradual adjustment after due notice; and where conflicting interests must be reconciled, the question will always be decided from the standpoint of the greatest good of the greatest number in the long run.

These general principles will govern in the protection and use of the water supply, in the disposal of timber and wood, in the use of the range, and in all other matters connected with the management of the reserves. They can be successfully applied only when the administration of each reserve is left very largely in the hands of the local officers, under the eye of thoroughly trained and competent inspectors.

These orders set forth the basic principles upon which the forest work of the nation is to-day carried on.

The new fiscal year, 1906, was noteworthy in accomplishment. The reserved areas were increased to 106,999,138 acres. Timber sales increased from 96,000,000 feet in 1905 to 288,-000,000 feet in 1906. Free use permits also increased. The Forest Service collected total revenues of $757,813.01 as compared with $73,276.15 in 1905. Of the former amount over $500,000 resulted from fees for grazing stock on forest preserves. The charging of fees for grazing constituted a new policy (though grazing regulations, without fee, had been in force for some years), and the resultant showing had its effect upon Congress, which limited the system of unrestricted expenditure of revenues in the forest reserve special fund (Act of June 30, 1906; 34 Stat. L., 669, 684), which had been permitted under the act of February 1, 1905. This limitation became effective July 1, 1908.

The new law also included a provision requiring that, beginning with the fiscal year 1906, ten per cent of the forest reserve revenues collected each year should be distributed pro rata for the benefit of public schools and public roads in the states and territories in which were located income-producing forest reserves. With the rapid expansion of the national forest area and the growth of receipts therefrom, western states and local communities in those states had complained that the reservation policy of preventing the acquisition of forest lands by private owners, tended to deprive them of an adequate basis of taxation. The grant of a share of the forest receipts was the answer to this complaint.

This provision again appeared in the act of March 4, 1907 (34 Stat. L., 1256, 1270), while the act of May 23, 1908 (35 Stat. L., 251, 260), increased the local share to 25 per cent and made the legislation permanent. The 1906 and 1907 clauses contained a stipulation that the amount paid to any county might not exceed 40 per cent of that county's revenue from all other sources.

National forest receipts increased to $1,530,321.88 in 1907, and in 1908 to $1,788,255.19.

Two other noteworthy acts were passed in 1906. One was the Forest Homestead Act, approved June 11, 1906 (34 Stat. L., 233), which permitted the agricultural use of lands within forest boundaries which were more valuable for crop raising than for timber growing, provided private ownership of such tracts should not interfere with the development of the forest as a whole. The other was the act of June 8, 1906, " for the preservation of American antiquities" (34 Stat. L., 225), or features of scientific and historical interest, located upon lands controlled by the United States, which provided that reservations or " monuments " created by Presidential proclamation within the boundaries of national forests were to be administered by the Forest Service. Fifteen such monuments have been created in national forests, with a total area of over 375,000 acres.

The act of March 4, 1907, abolished the forest reserve special fund (upon which the 1906 law had merely placed limitations) by providing that after July 1, 1907, all forest receipts should be covered into the Treasury as miscellaneous revenue. It also substituted the designation "national forests " for " reserves," and increased the salary of the Forester to $5000.[13] Finally, the new law provided that no more new national forests might be created or old ones enlarged in Oregon, Washington, Idaho, Montana, Colorado, and Wyoming except by act of Congress,[14] but the effect of this restriction was lessened by the action of President Roosevelt who, on March 2, two days before the act

[13] See also acts of July 14, 1890; March 1, 1899; March 2, 1901; June 3, 1902; and March 3, 1905.

[14] This restriction was extended to California by the act of August 24, 1912 (37 Stat. L., 497), and to Arizona and New Mexico by the act of June 15, 1926 (44 Stat. L., 745).

took effect, set aside twenty-one new reserves with an area of over 40,000,000 acres in the states affected.[75]

Two clauses of the appropriation act of May 23, 1908 (35 Stat. L., 251, 259, 260), permitted the inauguration of new administrative and fiscal policies. The first authorized the use of Forest Service funds, irrespective of the objects for which appropriated, for fighting forest fires in emergencies. The second began the system of permanent forest improvement which, with slight change, has continued in operation.

The year 1908 saw the general conservation movement at its height, and a series of important conferences on this subject were initiated, resulting more or less directly from the Public Lands Commission findings of 1903.[76]

The Forest Service had been engaged in certain coöperative activities beyond its own department, one involving supervision over the sale of timber on military reservations and a technical survey of such timber, and another being concerned with the management of the forests on a number of Indian reservations. The latter plan was now abrogated, and the management was taken over by the Office of Indian Affairs, under the act of March 3, 1909 (35 Stat. L., 783), which set aside $100,000:

To enable the Commissioner of Indian Affairs under the direction of the Secretary of the Interior to make investigations on Indian reservations and take measures for the purpose of preserving living and growing timber, and removing dead timber, standing or fallen; to advise the Indians as to the proper care of forests, and to conduct such timber operations and sales of timber as may be deemed advisable and provided for by law.

[75] There was still opposition to forestry in the West, of which this legislation was a reflection. The legislature of the State of Washington memorialized Congress on February 21, 1907, complaining of grazing " taxes," usurpation of the law-making power by the Secretary of Agriculture, " arbitrary edicts," and " petty exactions and pesterings by . . . officers and underlings." See Cong. Record, 59 Cong. 2 sess., pp. 3561, 3562.

[76] The first was a Conference of Governors summoned by President Roosevelt in March, 1908, after adjournment of which, the President appointed a National Conservation Commission, which compiled and submitted an exhaustive inventory of national resources in 1909. (60 Cong., S. doc. 676.) Following this a North American Conservation Conference was held at Washington.

A similar provision has been included in each Indian appropriation act since that time.[77]

On December 1, 1908, an administrative change of momentous nature was put into effect. This was the Pinchot decentralization policy, or, better, the system of localized national forest administration which was to prove a source of strength to the organization. It had been preceded by a rotation of service system under which supervisors of the forests were brought to Washington and placed in charge of pertinent work for terms of from two to three months, the purpose being to familiarize them with the problems involved in headquarters administration. With the new system six district offices, each under a district forester, were established in Missoula, Denver, Albuquerque, Ogden, San Francisco, and Portland, Oregon. District foresters were given certain local authority and were held responsible for results, though matters of larger administrative moment were required to be referred to Washington.

In July, 1909, decentralization was carried further by the establishment at Ogden, of a depot for the distribution of supplies and materials to the forest districts of the West and accounting records were simplified to reduce duplications, the Washington office keeping only general accounts of receipts and expenditures in the districts and leaving to the supervisors the details of transactions in the forests which they supervised.

This plan of organization, in its main features at least, persists to-day.

[77] See acts of March 28, 1908 (35 Stat. L., 51), and March 3, 1911 (36 Stat. L., 1076). The Indian Forestry Division, directed by technically trained foresters, administers, under approved forestry principles, the forest and grazing resources on sixty-four reservations in seventeen states, with an area of over forty-two million acres. The stand of timber is estimated at thirty-five billion feet. Timber sale receipts range from $2,500,000 to $2,800,000 annually and grazing receipts approximate $700,000 annually, or a total of from $3,200,000 to $3,500,000. Much timber and forage are used by the Indians for which no payment is made. The organization consists of a chief forester and ten assistants with general duties of inspection and administration, some thirty-nine trained foresters (all graduates of forest schools) in charge of various reservations, and 150 rangers, guards, scalers, and special employees.

4

In 1908 there was established the first experiment station designed to study the problems of the forest on the ground and at first hand; to accumulate scientific data upon which forestry procedure suited to American conditions might be based. This station was on the Coconino Plateau in Arizona. Other stations were soon established in Colorado, Idaho, Washington, California, and Utah—one station to each administrative district.

The necessity for laboratory research in forestry had been recognized by Fernow in the late eighties, and the idea had not been negelected. The laboratory work had progressed slowly and attained certain successes until, with the reorganization of 1905, the Office of Forest Products was established. At this time studies were undertaken with regard to the mechanical properties of selected woods, in coöperation with certain universities [78]—a method, by which, prior to 1910, the greater part of the Service research work was conducted.

In the fiscal year 1907, changes were made in the Office of Forest Products. Two sections (Timber Tests and Lumber Trade) were merged into one, and Wood Utilization and Dendro Chemistry became Wood Chemistry. Two new sections were added: Forest Measurements and Reserve Engineering. By 1909, the Office of Forest Products had become the Branch of Products, including the Offices of Wood Utilization, Wood Preservation, and Publication. In October, 1909, the Office of Wood Utilization was moved from Washington to Chicago because of the centering in the latter place of many directing offices of the lumbering and wood using industries.

Insufficient appropriations prevented the establishment of a research laboratory on an effective scale, so that efforts were made to establish one in coöperation with a university. An

[78] Purdue, Yale, California, Oregon, and Washington universities, and, later, Colorado University. Experiments were also conducted in a small laboratory in Washington and in a laboratory established at the Louisiana Purchase Exposition at St. Louis under a special appropriation. Several railroads also coöperated in practical tests of seasoning, handling, and treating tie and bridge timber.

arrangement was made with the University of Wisconsin, under
which the University furnished a suitable building with heat,
light, and power, while the Forest Service provided personnel
and contributed the raw materials. In June, 1910, the Forest
Products Laboratory was finally opened at Madison with a
staff of forty-five employees.

The early years of this enterprise were largely devoted to
the carrying on of the coöperative research work which had been
initiated in connection with various universities. As these
studies were brought to a close the work of the laboratory was
directed toward studies in timber mechanics, wood preservation, kiln drying, timber physics, paper pulp, and wood
derivatives.

Meanwhile the great initial era of the Forest Service, begun
in 1905, had come to a close with the dismissal of Gifford
Pinchot on January 7, 1910.[79] A new Forester, Henry S.
Graves, was appointed on January 12, 1910, effective
February 1.

Under the new régime numerous changes were inaugurated,
many of them looking toward a limitation of Forest Service
powers. All legal work of the Service was placed under the
" immediate supervision and direction of the Solicitor of the
Department of Agriculture" by order of the Secretary of Agriculture, and this action was reaffirmed by the act of May 26,
1910 (36 Stat. L., 416). By the same action the disbursing
and accounting work of the Forest Service was placed under
the supervision and direction of the Chief of the Division of
Accounts and Disbursements of the Department. The Secretary also assumed immediate control of all Service publicity.

On June 27 a joint order was signed by the Secretaries of the
Interior and Agriculture directing that thereafter all reports
by Forest Service officers on claims cases be sent to the Secre-

[79] Executive order No. 1142, November 29, 1909; Department of Agriculture, Annual Report, 1910, p. 891, and 61 Cong., S. doc. 719, p. 1297,
et seq.

tary of the Interior through the Secretary of Agriculture.[80] The purpose of this procedure was to enable the Secretary of Agriculture to check the reports and recommendations of the Forest Service. It was soon found that this practice resulted in much delay and that the check was unnecessary. Procedure was thereupon adopted whereby the District Foresters submitted their reports and recommendations directly to the field service of the General Land Office, with authority to the Chief of Field Division of the General Land Office to decide whether a contest should be initiated against the claim, with proper opportunity for appeal in behalf of the Department of Agriculture by the Solicitor of that Department. On August 5, 1915, further joint regulations were approved by the Secretaries of the Interior and Agriculture, under which the Department of Agriculture was recognized as the active contestant in all claims cases in which adverse reports were made and charges preferred by Forest Service officers.[81]

As pointed out earlier in this chapter, agitation for the extension of the forest system to the East had begun as early as 1899 through the efforts of the Appalachian National Park Association in North Carolina. The Bureau of Forestry, in 1901 examined nearly ten million acres in the Southern Appalachians, in coöperation with the Geological Survey, with a view to determining the suitability of the area for the contemplated reservations. Between 1901 and 1908 the question of creating national forests in the Appalachian region had fre-

[80] Under agreement between the Departments of Agriculture and the Interior, made in 1906, Forest Service supervisors made reports upon the validity of mining and agricultural claims arising within their forests. These reports were transmitted to the Department of the Interior and served as a basis in that Department for determining whether a contest should be initiated against the claim. In practice the General Land Office generally rejected a claim where an adverse report had been submitted by the Forest Service.

[81] Law officers of the Department of Agriculture were permitted to attend and participate in all hearings, and were accorded the right of appeal from the decisions of the General Land Office.—Department of Agriculture, Annual Report 1911, p. 785.

quently appeared, one manifestation of which was the act of March 4, 1907 (34 Stat. L., 1269), which provided for a survey by the Forest Service, of forest, land, and water conditions in the White Mountains as well as in the southern watersheds. As a result the Secretary of Agriculture in 1908, recommended the establishment of national forests in both regions.[82] In December of that year, also, the House Committee on Agriculture held extensive hearings on "Bills having for their object the acquisition of forest and other lands for the protection of watersheds and conservation of the navigability of navigable streams."

Debate ensued on the effect of forests on water-flow and rainfall. The House Committee on the Judiciary had reported that if forest reserves would aid navigation, Congress had the constitutional power to acquire such reserves,[83] but proof that forests would have such effect had not been sufficiently con vincing. The constitutionality of any act acquiring forests upon such grounds was questioned, therefore; nevertheless, the "Weeks Law" of March 1, 1911 (36 Stat. L., 961), was passed " To enable any state to coöperate with any other state or states, or with the United States, for the protection of the watersheds of navigable streams, and to appoint a commission for the acquisition of lands for the purpose of conserving the *navigability* of *navigable* rivers." [84]

The constitutionality of the new act was premised on the provision of the " Commerce Clause " of the Constitution and, since there seemed to be general acceptance of the necessity and desirability of the law, this construction was not questioned.[85]

[82] Annual Report 1908, pp. 433-34; also 60 Cong., S. doc. 91.

[83] 60 Cong., H. rep. 1514.

[84] Italics not in original. The " navigation " limitation upon the federal government had been given expression, previously, by the National Waterways Commission, created by the act of March 3, 1909 (35 Stat. L., 815, 818), which had said in its preliminary report (January 24, 1910): " It should always be borne in mind that the waterway improvements made by the Federal government, under the exercise of its authority should be restricted to navigation."—61 Cong., S. doc. 31, p. 20.

[85] It was argued that what makes for the navigability of rivers furthers commerce, that forests at the headwaters of rivers make them more

Apart from forest-river and forest-climate objects, the eastern national forests were a vital need for numerous reasons, among which were: to provide demonstration areas for scientific forest management, reforestation, etc.; to afford eastern national recreational areas and wild-life protection; and to provide future timber supplies for the East.

The new law was epochal in various ways. It established the principle of purchasing private lands to incorporate into national forests, it introduced in the national forest policy the plan of federal contribution to such state fire-suppression organizations as complied with the standards set up by the Forest Service, and it marked the final passing of governmental forestry out of the tentative into the fixed and definite.

Proposals covering a million and a quarter acres of selected lands had been received by June 30, 1911, and in November the initial purchase of eight thousand acres in North Carolina was made.[86]

The question of fire protection and suppression in the national forests received increased attention in this decade, 1910-20. A series of disastrous fires in the national forests in 1910 [87]

---

navigable and, that hence the purchase of forests about the upper reaches of navigable rivers was within the purview of the commerce clause. The necessity for the new legislation lay in the fact that in the states concerned there was no " public domain " and hence the national government could not set aside reserves.

The opposing viewpoints as to the effects of forests or deforestation on the navigability of rivers were maintained, and in 1910 the Weather Bureau and Forest Service initiated a coöperative experiment designed to throw light on the question and produce data on forest effect on climate, erosion, storage, etc. Though this experiment was conducted under the most careful observation for some seventeen years, it did not produce conclusive results.

[86] The selection of lands for purchase was made under authority of the National Forest Reservation Commission, set up under Section 4 of the new act, consisting of the Secretary of War, the Secretary of the Interior, the Secretary of Agriculture, and two members of the Senate, selected by the President of the Senate, and two members of the House of Representatives, selected by the Speaker.

[87] Seventy-eight fire fighters lost their lives and, in the national forests alone, more than $25,000,000 in timber was burned—some fifty times the normal. Some believe this situation was a vital factor in hastening the passage of the Weeks Act.

brought the Forest Service face to face, for the first time, with the size of its task, and showed the need for better preparation, greater resources, and extensive construction of improvements to equip the forests with means of communication and transportation. The necessary appropriations had to be obtained from Congress, together with legislative provisions that would give sufficient flexibility in the use of funds so that they could be applied in accordance with emergency needs. Personnel had to be trained, a smoothly functioning defensive organization elaborated, tested, and gradually improved, effective methods of fire fighting developed, and measures of prevention to lessen the number of fires studied and applied.

The act of June 4, 1897 (30 Stat. L., 11, 35), had directed the Secretary of the Interior to make provision " against destruction by fire " in the forest reserves, and acts in the following years, up to 1905, provided for " foresters and other emergency help in the prevention and extinguishment of forest fires." After 1905, the fire appropriations were made in the annual agricultural appropriation acts under " general expenses." These covered the national forests only and ignored fire suppression in timber on the unreserved public domain.

The Agricultural appropriation act of May 23, 1908 (35 Stat. L., 251, 259), making appropriations for the fiscal year 1909, contained a clause reading : " hereafter advances of money under any appropriation for the Forest Service may be made to the Forest Service and by authority of the Secretary of Agriculture to chiefs of field parties for fighting forest fires in emergency cases." The object of this legislation was to provide a means whereby temporary labor hired for fire fighting could be paid off at once in the field by having a " special deputy fiscal agent " on the spot with the necessary funds to disburse. Inasmuch as at that time a single lump sum appropriation provided for most of the expenditures of the Forest Service, not only for national forest administration and protection, but also for research and other activities, the " object for which appropriated " created no difficulties.

In consequence of criticism of the lump sum method of appropriation, which was called by some members of Congress a " slush fund," the Agricultural appropriation act of May 26, 1910 (36 Stat. L., 416, 424), making appropriations for the fiscal year 1911, split up the general item and made a specific appropriation for the administration of each one of the national forests individually, or considerably more than one hundred individual items. To offset the inflexibility resulting from this split up, however, a proviso was added permitting not more than ten per cent of all sums carried as subitems under " General Expenses, Forest Service," and also under the item " Improvement of the National Forests," to be used " in the discretion of the Secretary of Agriculture . . . . for all expenses necessary for the general administration of the Forest Service." The following year the leeway was increased to fifteen per cent. This interchangeability feature, however, was not confined in application to switching of funds in order to meet the cost of fighting forest fires.

Beginning with the fiscal year 1910 the Agricultural appropriation act carried an authorization of an interchange up to ten per cent by the Secretary of Agriculture " of the foregoing amounts for the miscellaneous expenses of the work of any bureau, division, or office herein provided for " (35 Stat. L., 1039, 1057). This was not a provision for switching between bureaus, nor did it apply to other items than those included with the general expenses of each bureau, but it added to the flexibility of Forest Service funds.

A separate item relating to forest fire control on the national forests first appeared in the appropriation act for the fiscal year 1911 (36 Stat. L., 416, 430): " For fighting forest fires and for other unforeseen emergencies, $135,000."

For the fiscal year 1912 (36 Stat. L., 1235, 1252), this item was increased to $150,000, of which sum $70,000 was made " immediately available," and an additional $1,000,000 was provided for the Secretary of Agriculture " for fighting and preventing forest fires in cases of extraordinary emergency."

The latter item dropped the following year to $200,000, while the item under the Forest Service portion of the bill continued at $150,000. The appearance of all these " extraordinary emergency " items was a direct result of the great fires of 1910. Gradually there was substituted, in place of an attempt to provide specifically, beforehand, a large emergency appropriation against the contingency of a bad fire year, the conception of meeting the emergency by drawing upon general expense funds to whatever extent might be necessary and looking to Congress to replenish these funds by a deficiency appropriation later in the year.

Various sums appropriated for permanent improvements were made interchangeably available in 1915, affording flexibility of funds for use in emergency fire conditions. In 1919 payment of rewards for the arrest of forest incendiaries was authorized. In 1920 coöperation with the War Department was established for airplane forest patrol during the fire season.

While these developments applied to the national forests only, coöperation among private forest owners for protection from fire progressed also, and the state coöperative work developed steadily during these years. Agreements for joint protective work between the Forest Service and the timber-owning railroads were entered into. Similar agreements were made with such organizations as the Oregon Forest Fire Association and the Washington Forest Fire Association, and similar associations in the white pine belt of northern Idaho and in the pine and redwood regions of California.

Protection against fire (and spoliation) had also claimed attention as early as August 1, 1914, in connection with certain railroad grants of timber land which had finally reverted to public control. An appropriation of $25,000 was made in 1914 (38 Stat. L., 644) for protection of timber on former railroad grants in Oregon and California, the work to be performed by the " Secretary of the Interior with the coöperation of the Secretary of Agriculture, or otherwise, as in his judgment might seem most advisable."

Pertinent to the fire protection work, also, was the Branch of Public Relations, established May 20, 1920, " to provide for better coördinated and supervised activities along this line and for more careful planning of methods by which public interest may be increased in both the protection and the use of the forests." [88]

Water Power. Meanwhile, questions regarding the use of water power in the forests had arisen. The act of May 14, 1896 (29 Stat. L., 120), was the earliest statute providing for water-power development on government land. It authorized the Secretary of the Interior to issue permits (but no easement, right, or interest) for the use of public lands for generating and distributing electric power. On May 11, 1898, an act was approved (30 Stat. L., 404) permitting rights of way over public lands for irrigation, to be used also for water-power development subsidiary to irrigation use. The act of February 15, 1901 (31 Stat. L., 790), like that of 1896, authorized the granting of revocable permits only, by the Secretary of the Interior at discretion, and this power (so far as the forest lands were concerned) passed, with the forests, to the jurisdiction of the Secretary of Agriculture under the act of February 1, 1905 (33 Stat. L., 628), and hence to the Forest Service.

Regulations under the act of February 15, 1901, were promulgated by the Secretary of the Interior on July 8, 1901, prescribing the manner of preparing and filing applications for permits, and the procedure thereafter necessary for obtaining the Secretary's approval. No formal permits were issued, the Secretary merely putting an appropriate endorsement on a map of the project submitted to him.

Under this act of 1901, provision was made for hydro-electric power development upon public lands and reservations under a revocable permit of land occupancy. The regulatory power thus conferred upon the administrative officers of the government first began to be exercised following the transfer of ad-

[88] Department of Agriculture, Annual Report, 1920, p. 229.

ministration of the national forests from the Interior to the Agricultural Department in February, 1905.

The transfer placed Forester Pinchot in charge of the forests. He moved at once to introduce the principle of fair compensation to the public for all commercial forms of use of national forest lands or resources. He initiated a Forest Service water-power policy of which the essential features were: (1) A charge for use of the land based on its value for power-development purposes, (2) the prevention of speculative acquisition of power sites or privilege of use unaccompanied by development, (3) the orderly and eventual full development of power resources as against partial utilization that would obstruct or prevent full use later, and (4) protection of the public interests through provisions that would make possible effective regulation of rates, services, and financing methods, and particularly to prevent inclusion in the rate base of unearned increment in land values or of inflated capitalization, however brought about.

The whole regulatory policy rested on land proprietorship. That mere revocable permits afforded water-power enterprises an inadequate basis for the development of projects on federal lands, was recognized by Pinchot and his associates, and they desired legislation that would allow long-term leases or contracts but without diminishing the power of the government as landowner to prescribe the conditions deemed necessary to the protection of the public interests.

The raising of the question of the need for federal regulation of water-power development involving the use of public land led to the broadening of the question to include water-power development on navigable streams. The authority of the government to impose conditions designed to safeguard and promote their navigability was fully established and was being exercised. Whether this authority extended far enough to afford a basis for applying a regulatory policy similar to that applied through federal land ownership was questioned. The conservationists held that it did. Others sought to prevent the application of such a' policy, maintaining that it would violate private property rights and the rights of the states and that it was uncalled for.

Authority to construct a dam affecting navigation had to be obtained through a special act of Congress and through approval of the construction plans by the War Department. The general dam act of June 21, 1906 (34 Stat. L., 386), authorized the War Department to impose such conditions as it might " deem necessary to protect the present and future interests of the United States." The question was raised whether under this authorization a charge could be imposed and a time limitation set—the two points on which at first the contest for regulation chiefly centered. In vetoing the Rainy River bill in 1908 President Roosevelt said:

All grounds for such doubt should be removed henceforth by the insertion in every act granting such a permit of words adequate to show that a time limit and a charge to be paid to the government are among the interests of the United States which should be protected through conditions and stipulations. . . . I do not believe that natural resources should be granted and held in an undeveloped condition either for speculative or other reasons. . . . We are now at the beginning of great development in water power. . . . Already the evils of monopoly are becoming manifest. . . . The present policy pursued in making these grants is unwise in giving away the property of the people in the flowing waters to individuals or organizations practically unknown, and granting in perpetuity these valuable privileges in advance of the formulation of definite plans as to their use. . . . In place of the present haphazard policy of permanently alienating valuable public property we should substitute a definite policy.

The veto message specified as requisite in such a policy: (1) A limited time within which plans must be developed and the project executed; (2) provision for annulling the grant if the conditions were not complied with; (3) provision to insure full development of the navigation and power or at least to prevent later obstruction of full development; (4) a license fee or charge so adjustable in the future as to secure control in the

interest of the public; (5) a definite time limit on the duration
of the grant. Thus a policy virtually identical with that of
the Forest Service with respect to power development on national
forest land was proposed for power development affecting the
navigability of streams.

The issue was clearly drawn between the advocates of federal
control and the advocates of private ownership of power sites.
Advocacy of federal control had been initiated as an organized
movement by Roosevelt's creation of the Inland Waterways
Commission early in 1907, with Pinchot as a member. In
appointing the Commission Roosevelt said:

The control of our navigable rivers lies with the federal
government, and carries with it corresponding opportunities
and obligations. . . . It is becoming clear that our streams
should be considered and conserved as great natural resources. . . . The time has come for merging local projects and
uses of the inland waters in a comprehensive plan designed for
the benefit of the entire country. Such a plan should consider
and include all the uses to which streams may be put.

A year later the Inland Waterways Commission submitted a
preliminary report, with findings and recommendations. It had,
however, reached the conclusion relatively early in its inquiry
that a conference on the general subject of the conservation of
natural resources was desirable, and in October, 1907, had
transmitted to the President a communication asking that he
call such a conference. This led to the so-called " White House
Conference of Governors," previously mentioned, which gave impetus to the broad movement for conservation as a national policy,
and resulted in the appointment of the National Conservation
Commission, with Gifford Pinchot as chairman. At a Joint
Conservation Conference, which was held in Washington in
December, 1908, the Conservation Commission submitted an
extensive report dealing with the natural resources of the
country under the four heads of minerals, lands, forests, and
waters. Both in the waters section of this report and in the

report of the Inland Waterways Commission the central thought was the need for a comprehensive treatment of each river system as a unit from source to mouth, with coördinated development of all uses. The necessity for improved facilities for water transportation assumed prominence and it was believed that co-ordinated power development might be made a source of revenue that would contribute substantially in time toward the cost of maintaining and extending navigability. Consequently, in the legislation enacted or debated for the next few years the question of a federal charge for the privilege of power development on navigable streams received more attention as a means for improving navigation than as a means for public control of power enterprises.

The necessity for a coördinated development of all forms of use and for the treatment of an entire river system as a single unit led to the search for some method of insuring a coördinated handling of the various executive and administrative functions of the government relating to water use.

Beginning in 1909 a study of water-power developments in the national forests was carried on by a specially employed hydro-electric engineer and conferences were held with representatives of power companies and the engineering profession. On the basis of data thus obtained, plus the experience of the Forest Service, a set of revised regulations was promulgated by the Secretary of Agriculture in December, 1910, after approval by the Forester and the Solicitor of the Department.[89] An act, approved June 25, 1910 (36 Stat. L., 847), authorized the President to withdraw from settlement, location, sale, or entry, any public lands or national forest lands, valuable for water-power development, and under this law the President made extensive withdrawals.

[89] The legislative history of control of water-power developments runs through the acts of May 14, 1896 (29 Stat. L., 120), May 11, 1898 (30 Stat. L., 404), February 15, 1901 (31 Stat. L., 790), February 1, 1905 (33 Stat. L., 628), May 1, 1906 (34 Stat. L., 163), March 4, 1911 (36 Stat. L., 1235), and December 19, 1913 (38 Stat. L., 242).

Although the water-power permits granted by the Secretary of Agriculture ordinarily ran for fifty years [90] and carried the privilege of renewal, they were revocable at pleasure, were not transferable, and in the last analysis granted no privilege but occupancy and use of forest land, terminable without notice. This, private capital felt, was too precarious an agreement upon which to invest large sums of money, and from 1908 on, bills had been introduced in Congress aiming to encompass security of investment and protection of the public interest.

In January, 1911, a joint committee was appointed by the Secretaries of Agriculture and the Interior, to study the problem and suggest legislation. A report was made in the same month which recommended the granting of absolute leases for a fixed period, irrevocable except for breach of conditions or of regulations later to be promulgated.[91] Interest in the subject continued, both in Congress and in the Department, and a report was issued in response to a Senate resolution of February 13, 1915.[92] Legislation was again urged by the Secretary of Agriculture in 1917 and 1918, but without result.[93]

During 1917, however, a new " Waterways Commission " was created,[94] the previous one having expired automatically in 1911. This act was made possible by a decision of the Supreme Court on March 9, 1917,[95] which disposed of the contention that there was no statutory or constitutional authority for national control of water-power development on forest lands, declared the act of 1901 the sole statutory authorization for commercial water-power development on national forest lands, and stated that such land could be held for such purpose only under permits, revocable at will, issued by the Secretary of Agriculture.

[90] Permits for shorter periods were sometimes granted. District foresters also issued permits for small, non-commercial developments.

[91] 62 Cong., S. doc. 274, p. 243.

[92] 64 Cong., S. doc. 316.

[93] Department of Agriculture, Annual Reports, 1917, p. 37, 1918, p. 42; also 65 Cong., H. rep. 715.

[94] Act of August 8, 1917 (40 Stat. L., 269). See act of June 10, 1920 (41 Stat. L., 1063).

[95] United States v. Utah Light and Power Co., 243 U. S. 389.

The necessity for legislation assuring capital and enterprise of sufficient permanency of occupancy to establish sound operations resulted in the act of June 10, 1920 (41 Stat. L., 1063), which established the Federal Power Commission. This measure had been first proposed by O. C. Merrill, while Chief Engineer of the Forest Service, who combined desirable features embodied in various bills, added important new proposals of his own, and provided an acceptable basis of agreement between the water-power interests and the conservationists. Under this act the permit-issuing power for water-power developments, even in the national forests, was given to the new commission composed of the Secretaries of War, Agriculture, and the Interior, and the Forest Service no longer had full control of power sites in its own sphere.[96]

Roads. One of the necessities for fire-fighting was trails or roads to make the forests accessible to the fire fighters, and this, coupled with the local community needs for road development, resulted in the initiation of a system of forest trails and roads.[97]

[96] This did not abrogate permits or operations under the old acts of 1901 and 1911 and the permittees might, if they chose, continue under these laws, subject to supervision of the Forest Service. It is to be noted, however, that, under the law above mentioned (Sec. 4 (b)), the Commission is authorized and empowered " to coöperate with the executive departments and other agencies of . . . the National Government " in water power investigations.

Since the Commission maintains only an Executive Secretary, it must depend upon other organization units of the government for service.

" As the Federal Power Commission cannot employ any persons other than the Executive Secretary, it has no field service. This makes coöperation with the field service of the departments all the more necessary. Particularly is this true in Alaska. . . ."—Institute for Government Research; Federal Power Commission, p. 72, Service Monograph 17 (1923).

The district foresters of the Forest Service represent the Federal Power Commission in the field, when necessary.

[97] Trails or roads were not built solely for fire protection, however, but for forest utilization as well; that is, for getting out timber, for recreational purposes, etc. Pertinent to the fire protection feature also was the act of March 4, 1911 (36 Stat. L., 1235, 1253), which provided $500,000 " for the construction and maintenance of roads, trails, bridges, fire-lanes, telephone lines, cabins, fences and other permanent improvements necessary for the proper and economical administration, protection and development of the National Forests."

As early as 1906 a highway engineer was detailed to study road and trail conditions in one of the forests with a view to planning improvements, and for five or six years sporadic work was carried on in extending and connecting trails, bridging streams, and enlisting the coöperation of local road officers, timbermen, and stockmen in such construction.

The act of August 10, 1912 (37 Stat. L., 269, 288), provided that ten per cent of all forest receipts on account of the fiscal year 1912 should be used for roads and trails, within the national forests in the states from which the receipts came.[98] This legislation was made permanent the next year by the act of March 4, 1913 (37 Stat. L., 828, 843). At first road and trail construction was carried on exclusively by the engineers of the Forest Service, but beginning in 1914 assistance was given by highway engineers from the Office of Public Roads.

In 1916 the scope of the forest road program was enlarged. The act of July 11, 1916 (39 Stat. L., 355, 358), appropriated $10,000,000 " for the survey, construction, and maintenance of roads and trails within or only partly within the national forests, when necessary for the use and development of resources upon which communities within and adjacent to the national forests are dependent." The funds were to be distributed over a period of ten years in equal installments, on an equitable coöperative basis to be determined by the Secretary of Agriculture and officers of the local coöperating governments.

Local expenditures were limited to ten per cent of the value "of the timber and forage resources which are or will be available for income upon the national forest lands within the respective county or counties wherein the roads or trails will be constructed," and the national government was to be reimbursed through annual ten per cent installments from the reve-

[98] Expended under the direction of the Secretary of Agriculture. The act of May 23, 1908, had allocated 25 per cent of forest receipts to counties of origin for roads and schools. The 10 per cent mentioned was additional. The same act underlay the process of "sales at cost" or non-commercial sales.

nues produced by the forests including, or adjacent to, the roads and trails constructed.

On September 1, 1916, regulations for administering this law were put into effect. They provided for the setting aside of ten per cent of each annual road appropriation for administrative expenses and apportioning the balance among the states concerned: one-half according to the proportion of the total national forest area in the state to the total area of the state and one-half according to the proportion of total estimated forest resources (both timber and forage) in the total national forest area in each state to the total value of similar resources in the entire national forest area of the nation.

The act of February 28, 1919 (40 Stat. L., 1189, 1200), was designed to improve coöperative relations and to obtain the funds essential thereto. The new law appropriated $9,000,000, in three yearly installments, for the same general purposes as were set forth in the act of 1916. Its language was broader, however, making direct mention of the upbuilding of forest administration, protection, and improvement and providing for direct construction, without local coöperation, of such roads and trails as the Secretary of Agriculture should consider necessary for the purpose above mentioned, or of national importance.

Meanwhile the act of March 4, 1915, was passed (38 Stat. L., 1095, 1100, 1101), permitting the leasing of small tracts in the national forests for summer home sites, hotels, resorts, and recreational purposes in general. This constituted the first major effort toward organized handling of the recreational facilities, which were soon to be recognized as forest resources comparable in economic importance with timber, forage, and water power. This had been anticipated by the act of February 28, 1899 (30 Stat. L., 908), but that law was of limited application and it had proved inadequate under the inundation of summer tourists and campers resulting from the development of the automobile.

Forest agricultural claims received further legislative attention in 1912. The act of August 10, of that year (37 Stat.

L., 269, 279), included certain provisions making it mandatory upon the Secretary of Agriculture to perform acts which had been discretionary under the Forest Homestead Act of June 11, 1906. The new provisions concerned the gradual elimination from Forest Service control of lands more valuable for agriculture than for other forest uses, and they were re-enacted in all succeeding agricultural appropriation acts to June 30, 1914, since which time the mandatory language has been omitted.

The act of August 10, 1912 (37 Stat. L., 269, 287), also directed the Secretary of Agriculture to sell at cost, to homestead settlers and farmers for their domestic use, the mature, dead, and down timber in the national forests.

During the early years of the period 1910-20 the Forest Service had attained, generally speaking, the form of organization which exists to-day. In 1911 there was a main office in Washington, organized administratively into six branches, under the general direction of the Forester and his Associate. Each branch handled one of the following functions: Operation, Grazing, Lands, Appalachian Work, Products, and Silviculture. This plan of organization was duplicated in each one of the six forest districts into which the country had been divided, and placed under the general direction of a district forester.[99]

This organization administered a total net forest area of over 168,000,000 acres, with a personnel of over 2600 employees and an appropriation of $5,938,100. The legal work, however, as well as the accounting, was cared for by the general departmental organization.

An important organization change came on June 1, 1915, when all research activities of the Service, except those relating to grazing, were merged in a new unit called the Branch of Research. This permitted the consummation of the policy of segregating investigation from administration and prepared the

[99] " Appalachian Work " was an exception to the district duplication.

way for the greatly increased war-work of the Forest Products Laboratory, which largely abandoned its peace-time programs during 1917 and 1918.

The Service Since 1920. Progress in the development of Forest Service activities and organization since 1920 has been continuous and substantial. The strengthening of coöperative relations, the consolidation of certain powers and duties by new legislation, and the expansion of existing activities—the working toward a definite national forest policy—have marked the period.

In the matter of organization the district system was augmented. District No. 7 had been created on July 1, 1914, with headquarters at Washington, for the more convenient administration of the new forests created in the White Mountains and the Appalachians under the Weeks Law, as well as those reserved out of the public domain in Florida, Arkansas, and Kansas.[1] In response to a demand for local autonomy in Alaska, District No. 8 was created on January 1, 1921. This was composed of the two national forests in the territory, and a resident district forester was placed in charge. A ninth district, called the Lake States District, and covering the national forests in Michigan, Minnesota, and Wisconsin, was created on January 1, 1929.

After a service of ten years, as Forester, Colonel Graves retired early in 1920 and Colonel William B. Greeley succeeded to the office on March 14, 1920. Colonel Greeley served until May, 1928, when he was succeeded by Major R. Y. Stuart, previously Assistant Forester in charge of Public Relations.

Protection. The act of May 11, 1922 (42 Stat. L., 507), made the Secretary of Agriculture solely responsible for the protection of timber on the Oregon and California land grants. As a matter of practice this had been done from the beginning, the appropriations being regularly transferred to the Forest Service, for use, at the request of the Interior Department.

---

[1] This last forest was abolished by executive order in 1915.

Akin to fire as a menace to forests, and an obstacle to conservation, were forest insects and blights. Reports had been made on insect ravages as early as 1885, and serious infestations occurred in the Black Hills region in 1905 and 1906. In more recent years inroads of bark beetles in southern Oregon and northern California, and in the Kaibab Forest in the Southwest had to be fought. In 1922 the loss from these beetles was reported as a billion and a half board feet of lumber, valued at three million dollars and spread over a million acres. In 1923 the Forester recommended the formation of an insect protective system modeled after the existing fire protection system and prepared to coöperate with private owners and managing interests.

Roads. By 1920 a question arose as to what proportion of the forest road fund should be expended for roads of primary forest importance and what for roads locally desired for tourist use. Legislation was needed to further the construction of both types and to draw a distinction between them. This came with the highway act of November 9, 1921 (42 Stat. L., 212), Section 23 of which made separate appropriations for roads of general public importance and those of primary forest importance in the form of a fund of $5,500,000 for forest development roads and one of $9,500,000 for forest highways. Coöperation by the Secretary of Agriculture with local governments was made permissive but not mandatory.

The first fund was to be expended primarily for the welfare of the forest itself and the protection and economic handling thereof, and for these purposes $2,500,000 was made immediately available and $3,000,000 more on July 1, 1922. These amounts were to be apportioned to the various states and to Alaska and to Porto Rico, on a basis of comparative local forest need, to be determined by the Secretary of Agriculture. The second fund ($2,500,000 of which was made immediately available and the balance at the beginning of the following fiscal year) was allotted in a substantially similar manner, but was to be expended for highway rather than forest development.

This dual plan of development has been continued by authorization and appropriations in various postal and agricultural appropriation acts.

The need for these developments as an aid to fire protection was emphasized by fire losses during the years 1917-1921, in which period the annual loss from burning of national forest timber was $1,500,000 and the annual suppression bill $1,200,000.

Research. As has previously been mentioned, an early phase of forest investigative work was that of conducting experiment stations. At first these stations were developed wholly locally, but it became apparent that, for adequate service, they would have to be conducted on a regional basis for the benefit of all the forests in each region, both public and private. In the continental United States, it was recognized that there were twelve well-defined forest regions, each with conditions and problems peculiar to itself. In 1922 two experiment stations were established on the regional basis and nine more have since been created.

Grazing. In this decade the matter of grazing fees was a subject of controversy. Regulation of national forest pasturage began in 1897. In 1906 fees for grazing were first required and litigation followed. In 1911 the United States Supreme Court rendered two decisions in cases involving practically every phase of governmental range regulation.[2] In these two suits the constitutionality of Section 24 of the act of 1891 and of the act of 1897, providing for the protection of the forest reserves, was called in question. The government was upheld at all points, but this did not end opposition.

The first grazing fees, were low,[3] yielding little more than the cost of administration. Small local increases were made in 1910 and a small general increase in 1912. In 1915-16 and 1917-19

[2] United States v. Grimaud *et al.,* 220 U. S. 506; United States v. Light, 220 U. S. 523.

[3] From three to four cents per month per head for cattle and one and a quarter to two cents for sheep.

substantial increases were initiated[4] which produced, roughly, some three times the cost of grazing administration. In 1924 an elaborate appraisal of forest ranges was completed by C. E. Rachford. This involved a comparison with some twelve hundred tracts of state and private grazing land in the vicinity of the national forests, and the formulation of a program which would eliminate the flat rate and substitute rates for each pasturage, representing the value thereof in the light of accessibility.

The year 1921, however, had produced adverse conditions. Agricultural enterprises were suffering from post-war deflation and the financing and marketing phases of the livestock industry were hard hit. In addition, a succession of severe drought years had occurred. As a consequence emergency measures were necessary, specifically, the acts of March 3 and August 24, 1921 (41 Stat. L., 1325 and 42 Stat. L., 186),[5] which postponed the payment of grazing fees (ordinarily due thirty days before the opening of the grazing season—about March 1) to September 1 and later to December 31. In spite of this, many permittees became delinquent,[6] and the prospective increase of rates was protested by the American National Live Stock Association. The Secretary of Agriculture promised that there would be no increase in rates in 1924 or 1925, and the regulations were somewhat modified by the joint resolution of March 3, 1925 (43 Stat. L., 1259), but the stockmen, in convention, adopted a resolution protesting against fees greater than necessary to cover cost of administration and urged legislation to this effect.[7]

By 1925 irritation had been allayed by new regulations which instituted permit terms of ten years,[8] the requirement of proper equipment, and exemption limits protecting permittees from

[4] To about eleven cents (average) for cattle and three cents for sheep.
[5] See also joint resolution of November 17, 1921 (42 Stat. L., 220).
[6] For three years, 1922-24, delinquencies amounted to $126,000, mostly uncollectible. The total grazing fees for these years were about $7,000,000.
[7] See 68 Cong. 1 sess., S. 2424, and 68 Cong. 2 sess., S. J. R. 147.
[8] Five-year permits had been granted on some forests in 1919.

future reductions in number of stock [9] below an economic herd as determined by local custom. These regulations went into effect March 1, 1924, except the ten-year permits, which began from January 1, 1925.

On March 4, 1925, the Senate adopted a resolution [10] authorizing the Committee on Public Lands and Surveys to make an investigation into the entire question of national forest and public domain administration. The hearings, as a matter of fact, gave major attention to the question of grazing administration and control on the forests and public domain and resulted in the introduction of bills looking toward adjustment of points in dispute.

Meanwhile, on December 30, 1925, the Secretary of Agriculture announced the remission of grazing fees,[11] for the year 1926, on the forest ranges of certain drouth-stricken areas in the Southwest. On January 18, 1926, further grazing regulation modifications were announced. The ten-year permits were given the full force and effect of contracts, irrevocable except for violation of terms. They contained provisions for possible herd reductions by forest officers during the life of the permit, but these were definitely limited in amount. The new regulations also permitted the encouragement of individual allotments of acreage, where practicable, and authorized suspension of additional grazing privileges [12] for stated periods in regions where economic production of livestock and use of ranch lands dependent upon forest grazing lands would be benefited thereby.

Finally the regulations provided for local grazing boards consisting of one member representing the Forest Service and

[9] Many complaints had been based on the instability of range tenure, that is, the fact that permits were on a year-to-year basis and also on the fact that it was uncertain, each year, how many head of cattle the forest officers would permit to run.

[10] 68 Cong. 2 sess., S. R. 347.

[11] Under authority of the acts of 1891, 1897, and 1905 and by virtue of the regulations under which grazing permits were issued by the Forest Service. Previous remissions had been made by special congressional authorization.

[12] That is the admission to the range of new permittees with the resultant cutting-down of the headage allowed the old.

others selected by the grazing permittees, to study and settle grazing questions and assist forest officers in the development of grazing policies, with jurisdiction either in single forests or groups of forests, depending upon conditions. Board decisions were to be final unless appealed to the Secretary of Agriculture, who had power of review and final decision.

In July, 1926, the stockman arbiter, appointed by the Secretary of Agriculture to review the Rachford grazing report of 1924, submitted his findings, in view of which a recheck of the original appraisal was made and a downward revision of rates was submitted in November. The arbiter recommended still further reduction, however, and his figures were approved by the Forester.[13] Decision was finally made by the Secretary of Agriculture on January 25, 1927, to the effect that the 1919 scale of fees would remain in force until 1928, that the recommended schedule would be put into effect on a graduated scale beginning in 1928 and extending to 1931, that the full rates would be applied from 1931 to 1934 inclusive, and that there would be no changes in the schedule during the ten-year period beginning in 1935 "unless there should be a material change in conditions."[14]

Recent Legislation. An act of March 20, 1922 (42 Stat. L., 465), bore superficial resemblance to the old forest lieu clause,[15] which had occasioned considerable trouble. It authorized exchanges of privately owned forest lands lying within the boundaries of national forests, for government owned land or stumpage in any national forest in the same state. Authority to act for the government was vested in the Secretary of Agriculture and the Secretary of the Interior. Where the old act gave a right, the new granted a privilege. The old act allowed selec-

[13] Increases, instead of 100 per cent, became 50. The average increases for cattle approximated four cents per head per month; that is, from 10.4 cents to 14.4 cents and for sheep from 2.9 cents to 4.5 cents.

[14] Lumbermen have always had to purchase stumpage on the basis of the highest bid.

[15] In act of June 4, 1897. Repealed March 3, 1905.

tions to be made by the private owner and prospective exchanger, of an equal *area* anywhere on the public domain. The new law permitted a trade for government holdings equal in *value,* regardless of area, *provided* the government cared to make the exchange; that is, the mandatory feature was eliminated.

The process of exchange was simplified. In case a trade was mutually desirable and the details satisfactory, the signing of papers was sufficient, no additional legislation being necessary.[16]

In 1923 but a few minor exchanges were made under the new act. Since then, the volume of business has steadily increased, resulting in a net area increase of 353,735 acres to the end of the fiscal year 1929. During these years the forests have acquired additional areas, however, in other ways than through trading, formal setting aside, or proclamation. For example, on May 21, 1923, the Supreme Court[17] declared fraudulent twenty-four timber and stone entries which had been patented in 1902, and such of this area as lay within a national forest reverted thereto with the vacation of the patents, while the other part was added by act of Congress.

Previous to 1920, under the act of May 11, 1908 (35 Stat. L., 102), the regulation of the killing of sea-lions and walruses in Alaska was under the jurisdiction of the Department of Agriculture, and of minks, martens, sables, and other land, fur-bearing animals, under the Department of Commerce. The act of May 31, 1920 (41 Stat. L., 716), adjusted the powers of each department to its functional and special equipment, but the act of January 13, 1925 (43 Stat. L., 739), superseded all previous laws relating to Alaskan game animals and birds, and land fur-bearing animals, and put practically all power in the hands of the Secretary of Agriculture through an "Alaska Game Commission," which he appoints.[18]

---

[16] An amendment to the new act was approved on February 28, 1925 (43 Stat. L., 1090), allowing the government and the private owner to make reservations of timber, minerals, easements, etc., provided all rights, etc., so affected, remain subject to state laws.

[17] United States v. Curtis, Collins and Holbrook, 262 U. S. 215.

[18] The Forest Service has been continuously concerned with game protection and administration since 1905, partly in coöperation with states

The act of May 11, 1922 (42 Stat. L., 507), carried a new item: an appropriation of $10,000 for the construction of camping facilities on the national forests, to care for the steadily increasing demand of automobile tourists. Such appropriations have since continued and increased.

On June 7, 1924, the Clarke-McNary Act was approved (43 Stat. L., 653), providing, in the words of the Forester, " the greatest step forward in forestry in America since the Weeks Act of 1911." The new law, first of all, made possible a great impetus in the all-important matter of forest fire protection, especially in the way of coöperation and state aid. It authorized an annual appropriation, for fire protection assistance to the states (and the development of farm forestry), of $2,700,-000. The act of February 10, 1925 (43 Stat. L., 822), appropriated $760,000 for these purposes for the fiscal year 1926, this in addition to the amounts appropriated under the Weeks Act and the regular Forest Service appropriation.

The law further, looking toward the encouragement of forest-growing, provided for the study of forest land taxation methods in the states and for assistance to farmers in forest planting and timber growing through federal and state coöperation. It provided means for recommending to Congress, for addition to the national forests, suitable portions of the public domain. It also authorized the President to establish, as national forests, suitable portions of other government reservations.

Study of the tax problem was begun in 1925 by a forest tax investigation staff, organized in New Haven by Professor Fred R. Fairchild. This staff moved to Minnesota for field study in 1926. The work was gradually extended to cover the entire Lake States region.

Section 6 of the new law permitted governmental acquisition of forest lands throughout larger and more important areas by

---

and other government agencies. The act of March 3, 1905 (33 Stat. L., 861), provided for coöperation of forest officers with local officers in game protection. The Forest Service, in 1912, procured the first dependable figures on the number of big game animals killed in the national forests. Until 1916, it carried on the extermination of predatory animals in the national forests as part of its regular work.

making the limitation on purchase areas " the water-sheds of navigable streams " instead of, as in the Weeks Act, the *head-waters* of navigable streams, and permitting purchases for " timber production" as well as protection of navigation. This constituted direct action toward a government program of reforestation.[19]

Most important of all, however, the Clarke-McNary Act ıepresented a forward step in the development of a forest policy for the United States and did so by coöperation rather than compulsion.

In November, 1924, the National Conference on the Utilization of Forest Products met in Washington, with the purpose of creating a national clearing house of facts looking toward the better utilization of the timber supply. The National Committee on Wood Utilization held its first meeting in May, 1925, and took steps to foster a national movement to save timber and eliminate waste by better methods of manufacture and use.

On January 26, 1927, the Chief Coördinator of the Bureau of the Budget established the Forest Protection Board, composed of representatives of the National Park Service, the General Land Office, the Office of Indian Affairs, the Bureau of Biological Survey, the Weather Bureau, and the Forest Service, "to formulate and recommend . . . general policies and plans for the protection of the forests of the country, especially in the prevention and suppression of forest fires, embracing measures for the coöperation of Federal, State and private agencies."

With regard to this board the following is pertinent:

Following a series of 11 meetings, the board presented a report to the chief coördinator which sets forth the action necessary in its judgment to coördinate the protection of the forest lands administered by the several Federal bureaus and to effect

[19] The constitutionality of this act, like the Weeks Act, with regard to the stream-flow feature, is questioned by many. General agreement on the merit and need of these laws, however, has probably prevented the question being brought to issue.

more adequate protection of State and private forest lands under the coöperative policy laid down in the Clarke-McNary Act. This is probably the first occasion on which a coördinated picture of the forest-protection problem of all of the lands in the continental United States, under whatever jurisdiction or ownership, has been presented in a single statement.

It is the purpose of the board to continue its collaborations with a view to working out more complete coöperation and coordination in the protective work of the Federal agencies and to make mutually available the results secured in the development of protection methods.[20]

Early in 1928 the board was enlarged by including representatives of the bureaus dealing with forest problems relating to insects, rodents, and tree diseases: the Bureaus of Entomology and Plant Industry.

In January of 1928 the board organized local committees to act for it, so far as possible, in handling local problems of interbureau coöperation in forest protection, etc. These were appointed from the field organizations of the eight bureaus, one committee in each of the six divisions of the West.

On April 30, 1928 (45 Stat. L., 468), the Woodruff-McNary Act was approved. This act authorized a series of annual appropriations to the aggregate of $8,000,000, to carry out the provisions of Section 7 of the Weeks Act, for the protection of the watersheds of navigable rivers.

Finally on May 22, 1928, came an act (45 Stat. L., 699) of broad significance and liberal provisions. The purpose was " To insure adequate supplies of timber and other forest products for the people of the United States, to promote the full use for timber growing and other purposes of forest lands in the United States, including farm wood lots and those abandoned areas not suitable for agricultural production, and to secure the correlation and the most economical conduct of forest research in the Department of Agriculture, through research in reforestation, timber growing, protection, utilization, forest economics, and related subjects, and for other purposes."

[20] Forester, Annual Report, 1927, p. 20.

. . . the Secretary of Agriculture is hereby authorized and directed to conduct such investigations, experiments, and tests as he may deem necessary . . . in order to determine, demonstrate, and promulgate the best methods of reforestation and of growing, managing, and utilizing timber, forage, and other forest products, of maintaining favorable conditions of water flow and the prevention of erosion, of protecting timber and other forest growth from fire, insects, disease, or other harmful agencies, of obtaining the fullest and most effective use of forest lands, and to determine and promulgate the economic considerations which should underlie the establishment of sound policies for the management of forest land and the utilization of forest products: *Provided,* That in carrying out the provisions of this Act the Secretary of Agriculture may coöperate with individuals and public and private agencies, organizations, and institutions, and, in connection with the collection, investigation, and tests of foreign woods, he may also coöperate with individuals and public and private agencies, organizations, and institutions in other countries; and receive money contributions from coöperators under such conditions as he may impose, such contributions to be covered into the Treasury as a special fund which is hereby appropriated and made available until expended as the Secretary of Agriculture may direct, for use in conducting the activities authorized by this Act, and in making refunds to contributors. . . .

This act, known as the " McSweeney-McNary Act," established a ten-year program of forest research, and authorized, within that period, a series of appropriations increasing each year until the maximum annual amount of $3,500,000 is reached.

The attempt has been made, in this chapter, to sketch briefly the development of the forest policy and national forestry organization and activities in the United States from the Revolution onward. It has been necessary, however, to stress those events which have had a more or less direct bearing on present organization and activities and, to that extent, pass over briefly certain matters which are, absolutely considered, of considerable moment in the history of the forest problem itself.

It is pertinent, now, to turn to a description of the activities with which the present Forest Service is concerned.

CHAPTER II

ACTIVITIES

" The Forest Service seeks to promote the best use, in the permanent public interest of all forest lands and forest products in the United States." [1]

In order to carry out this purpose numerous activities have been developed under permissive legislation and departmental order. The act of March 3, 1891 (26 Stat. L., 1095, 1103), stated:

That the president of the United States may, from time to time, set apart and reserve, in any State or Territory having public land bearing forests, *in* [*sic*] any part of the public lands wholly or in part covered with timber or undergrowth, whether of commercial value or not, as public reservations. . . .

This act gave the President discretionary power to create forests, but it was followed by the act of June 4, 1897 (30 Stat. L., 11, 35), which imposed certain limitations:

No public forest reservation shall be established, except to improve and protect the forest within the reservation, or for the purpose of securing favorable conditions of water flows, and to furnish a continuous supply of timber for the use and necessities of citizens of the United States; but it is not the purpose or intent of these provisions or of the Act providing for such reservations, to authorize the inclusion therein of lands more valuable for the mineral therein, or for agricultural purposes, than for forest purposes.

Subsequent paragraphs laid down the requirements regarding administration, which included protection of the forests from fire and depredations, and authorized the Secretary of the Interior to

[1] Col. W. B. Greeley in *U. S. Daily,* March 5, 1927.

. . . make such rules and regulations and establish such service as will insure the objects of such reservations, namely, to
regulate their occupancy and use and to preserve the forests
thereon [that is the reservations] from destruction; . . .

The first determination of the objectives of "such rules and
regulations," which constitute, in effect, the basis of the national
forest activity, came with a letter of the Secretary of Agriculture on February 1, 1905, reproduced in the preceding chapter.[2]
In that letter certain matters of policy are presented in bold
relief. These are:

. . . all land is to be devoted to its most productive use for the
permanent good of the whole people and not for the temporary
benefit of individuals and companies. . . . You will see to it
that the water, wood and forage of the reserves are conserved and
wisely used for the benefit of the home builder first of all. . . .
The dominant industry will be considered first, but with as little
restriction to minor industries as may be possible. . . . Where
conflicting interests must be reconciled, the question will always
be decided from the standpoint of the greatest good of the greatest
number in the long run.

These, then, are the fundamental laws, principles, and pro-
nouncements upon which all national forest activities are based.

The administrative work of the Service, in the broadest sense,
consists of formulating policies, making plans, and issuing orders. The specific activities in which the Service is engaged
may be conveniently grouped for discussion, under the following heads:

1. Utilization
2. Protection
3. Development
4. Investigation
5. Coöperation
6. Information
7. Acquisition

[2] See page 33.

Utilization. One of the most important technical duties of the Forest Service is to regulate the cutting of timber in the national forests so as to insure "a continuous supply of timber for the use and necessities of citizens of the United States." The timber disposal policy of the Service has been stated as follows:

The Government is not a landlord owner, but a trustee. As a trustee it must treat all alike and refuse permission to the first comer to pocket the share of all the rest. Hence timber is given away through free use permits only in small quantities to the actual home maker, who comes to develop the country and in larger quantities to communities for public purposes. Otherwise it is sold to the highest bidder, but under such restrictions as look to the maintenance of a lasting supply answering to the needs of the locality, to be had without favoritism and without extortionate demand based upon the necessity of the consumer.[3]

Management Plans. In order to work out this problem the Service has adopted a policy of dividing the national forests into units of management known as "working circles," upon which studies and inventories of the existing forest resources are made in order to gain some idea of the volume and area of timber of different ages and of the rate of growth. From the data thus obtained, taking into consideration the cutting methods employed for the type of timber on the area, there is then worked out an annual cutting rate that will maintain the proper balance between use and new growth. Speaking upon the subject of national forest management, a former Forester said:

Our management plans are based upon . . . general policies rather than a strict adherence to any theory of regulation. Fine spun calculations of sustained yield are not possible from the data to be had nor are they regarded as necessary at this stage in the development of the National Forests. . . . The problem presented by our great overstocking of timber in all stages of maturity and overmaturity can not be worked out by [a detailed] type of regulation. Nor is the data on growth and yield yet sufficient to justify it. We must deal with our man-

[3] Secretary of Agriculture, Annual Report, 1907, pp. 61-62.

6

agement problem along broader lines and in more simple and workable terms. The time for refined regulation of the European type will doubtless come, but only after we have worked over our forests into more like a normal distribution of age classes and also after much more comprehensive growth and yield figures have been secured.

In the present pioneer stage we are seeking rather to develop rough and ready plans of forest management, based primarily upon removing the stock of old growth timber and extending the period of its removal over the time that must elapse until the existing stands of young growth will become merchantable. 'The main purpose in most of these plans is to avoid a hiatus in timber cut after the old stock has been taken off and a disorganization of industrial and social institutions which depend on a continuous output from a given unit. In some cases this principle can not be adhered to strictly because of the silvicultural condition of the old timber or local difficulties in securing its operation at a rate which will permit a sustained yield when the change comes from the old growth to young stands. In many instances all that can be attempted is to control the rate of cutting so that we will have enough old timber to maintain operations until the young stands reach maturity, at which time a revision of the annual yield will be necessary. A scheme of logging under which the areas of old timber will be worked in the order dictated by their silvicultural condition, just as far as existing transportation facilities will permit, is always one of the most important features of the management plan.

The working circle, or unit of management in the National Forests, is laid out in accordance with topography primarily, or a combination of topography with transportation facilities from the timber to the nearest point for its manufacture. In a typical western National Forest, this usually means that each major watershed is a separate working circle. The boundaries of the circle are often adjusted to the specific industrial or community considerations which are placed among the purposes to be served by the management of the unit. This is particularly true in the small working circles on a number of eastern National Forests, where maintaining established local industries or providing woods labor for rural communities are set forth as definite objects of management. We believe that as far as practicable the government should get down to the specific conditions and needs of the locality in laying out its schemes of management; and this factor has been given more and more weight as our general policy has developed.

* * * *

It goes without saying that forest management on the National Forests is still in the rough and pioneer stage. We have an immense variety of conditions with which to deal. We are under the necessity of developing technical methods as we go. We can not withdraw these forest areas from use until schemes of regulation satisfactory to the technician have been completed; nor in our judgment is there need for such types of regulation under present conditions. Taking the two ideals of a sustained yield of timber products and a permanent and desirable type of forest industry and forest community, we believe that we should go right ahead under rough and ready plans of management dictated by common sense and by such growth and yield data as can be obtained. Constant improvement in these first management plans will of course, be essential. The whole scheme is fundamental to the conception of the national forests. It is simply a question of using the best tool that we can devise for immediate requirements and then constantly seeking to improve that tool.[4]

In the national forests there is a total timberland area of about 80,000,000 acres. This area contains between six hundred and seven hundred working circles. To January 1, 1930, something more than a hundred management plans had been worked out and approved for various of these circles; about thirty of them are for areas in the eastern or Weeks Law section of the country, and the others are scattered here and there in other regions, where they have been called into existence either by local demand for national forest timber, or for silvicultural reasons. Such plans are continually being prepared and new working circles placed under management or equipped with management plans.

This work has a triple aspect. There is, first, the formulation of the general policy by the responsible heads of the Service.

[4] *Journal of Forestry*, March, 1925, p. 223 *et seq.*

Next, in general conformity with this policy, there is the determination by the local officers (primarily those in charge of the forests with the approval of the district officers), as to how the general plan can best be applied locally. Finally, there is the direct formulation of the actual plans by the local staff in each forest, on the basis of available data on timber volume, growth, and yield. The forest officers are then in position to proceed, fortified by the knowledge of how much timber to cut, and when, where and how to cut it. After a plan is formulated by the local forest personnel, it must receive, first, district, then Service, approval.

This local determinative work is carried on by the local forest employees as a part of their regular work, though outside assistance is used in the case of important projects. The Service maintains a small force of loggers, lumbermen, cruisers, scalers, and logging engineers available for duty as, and where, required, but the bulk of the work is done locally. Plans are made for particular areas as economic or silvicultural considerations dictate.

The actual cutting and disposition of the timber is usually carried out by the persons to whom it is sold or given. Generally speaking, the Forest Service cuts no timber.

Sales: Advertised. There are consummated each year something like 13,500 sales of national forest timber. These sales, which may be classified as advertised, unadvertised, and " at cost " are made under the timber sale provisions of the 1897 law, as amended by the acts of June 6, 1900 (31 Stat. L., 661), and March 3, 1925 (43 Stat. L., 1132). All commercial sales of timber involving a stumpage value in excess of $500 are advertised. They constitute something like one-tenth of the total number made. Smaller sales may be, and sometimes are, advertised.

All large sales are governed by substantially the same basic conditions; namely, that the purchaser may remove, annually, not more than the amount of timber specified in the management plan governing the area in question. The price is readjusted at agreed-upon intervals during the progress of the operation, to conform with local rates, accessibility, cost of operation, etc.

The installation of permanent improvements is not unusual. Most of the large-scale sales of national forest timber in the continental United States are on mountainous and inaccessible areas that can only be utilized by the construction of lines of railroad for hauling the lumber out. The construction of mill buildings, fully equipped, is also a usual accompaniment of the larger timber sales.

Although large-scale operations attract the attention there are numerous sales of moderate size, while the greatest number of advertised sales are those involving small sums of money.

Sales: Unadvertised. Small sales made without the formality of advertising for bids are arranged by agreement between the purchaser and either the ranger in charge of that portion of the particular forest in which the desired timber stands or the forest officer in charge of the entire forest. When arranged through the subordinate officer, the sale is usually consummated in from one to five days; when the higher officer handles it the time required varies from one to fifteen days. Consummation of an advertised sale requires a longer period, since all advertisements must run for at least thirty days, and it occasionally takes as long as two hundred days to complete large-scale sales.

The sales thus far described, both informal and advertised, are known collectively as commercial sales. The timber in all such sales is sought by the purchaser for profit-making purposes and is sold for not less than its appraised value. This value is determined by consideration not only of quality and quantity of timber and of market prices for material of similar grade but also of accessibility and necessary preliminary outlay for roads, logging equipment, etc., to get the logs out.

Sales at Cost. In addition to the commercial sale there is also a cost, or non-commercial, sale, made only to homestead settlers and farmers for domestic use on their own holdings. Timber which can be cut and removed without injury to the forest is disposed of in sales of this kind. No direct charge is made to the purchasers for the timber itself, but they are required to pay the actual cost of making and administering the sale. Some 16,000,000 feet are disposed of annually in sales of this sort to about six thousand settlers and farmers. These transactions involve about one and one-half per cent of the total cut.

Removal Under Use. Besides disposition by sale, national forest timber may be removed under what is known as "free use," whereby timber in amounts not exceeding twenty dollars in value may be taken annually from the forest by individual users; or under "administrative use," whereby the Forest Service may dispose of varying amounts for the protection or improvement of the forests.

Free use has for one of its purposes, the protection and silvicultural improvement of the forest, restricted to bona fide settlers, miners, residents, and mineral prospectors for local use, to persons constructing private telephone lines across the forests which will be of value in protecting the forests from fire, and to certain branches of the national government for use in the prosecution of their particular lines of work. About 35,000 permits for free use are granted annually.

Administrative use is a phase of forest regulation, designed to cover all features of national forest timber utilization not covered by the ordinary sale or use permit. It may be exercised to protect the forest from injury, or to conduct investigations. Disposal may be accomplished by sale, free use, or otherwise, whichever may be most advantageous to the United States.

Utilization Procedure. Such details of utilizational procedure as emergency sales, which are sometimes made without regard to the ordinary procedure, sales of miscellaneous forest products, such as Christmas trees and various kinds of barks used commercially, and sales of naval stores in general, are governed by procedures substantially similar to those already outlined. The same general practice covers what is known as "timber settlement"; that is, the payment or other satisfaction for timber destroyed or removed in connection with the occupation of national forest land under permits of various sorts.

The refinements of the several procedures are available in various Forest Service publications, especially The Use Book, which contains, in elaborate detail, the various steps necessary to be taken in the procurement of national forest timber, or in the utilization of any other national forest resources, together with the regulations governing the same, made under the general authority of the 1897 law.

The cutting permits of the Service, which are contracts in effect, lay down the following requirements:

1. Location and description of the material purchased including the restriction of cutting to timber which has been marked or designated by a forest officer, and the basis for determining its merchantability
2. The price to be paid therefor, with provision for readjustment at intervals if the sale runs to exceed a certain period
3. The cutting time limit and minimum or maximum annual or periodical cut
4. The determination of the volume of the material by scaling, counting, measuring, or other methods
5. The protection of unsold timber during the sale and its future safeguarding through disposal of logging debris, etc.
6. The protection of national forest interests by the adoption and observance of an approved logging plan for the specific operation
7. The agreement to assist in preventing and fighting forest fires during the life of the operation.

The foregoing requirements are general. In addition there are special requirements applicable to particular cases.

Direct service activities in connection with the cutting of national forest timber have already been touched upon in what has been said of the staff work performed in the preparation of the management plans, based on the direct field work in the forests by the regular personnel, assisted by the roving force of specialists in cruising and estimating the timber, mapping the various age classes, computing the growth rate, and determining the rate of annual cut.

Even when the way has thus been cleared for the actual disposition of the timber, there is still much to be done by the Service. In the case of commercial sales there must be cruises and appraisals of the desired timber preliminary to the final transactions. These are generally made by the local forest force, but sometimes by the small special force already referred to, depending somewhat upon the size of the contemplated operation but more upon the circumstances peculiar to it. After the timber is appraised the conditions to be made a part of the contract must be determined upon in order to insure the carrying out of the Service requirements regarding silviculture, utilization, fire protection, etc. In the case of cost sales the cost of sale administration is determined by regional averages.

After a sale has been entered upon or a use permit granted, and before the cutting or removal begins, the timber to be taken must be marked or designated by the local forest officers.

When cutting actually begins, the work of the forest officers becomes supervisory in character. Operations must be watched in order to make sure that the Service requirements in silviculture and utilization are being complied with, that slash is being piled according to agreement, stumps cut at the proper height, young growth given all practicable protection, etc.

The final activity of the Service in connection with timber utilization is performed when the timber is taken out of the cutting areas and piled at convenient places for a final checking-up as to the amount taken. This is done by scaling, measuring, or counting the cut timber. In contract sales no timber may be removed from the place designated for scaling until it has been verified and marked by a forest officer. In the case of the use permits this requirement is sometimes waived when there is cause for believing that reasonable compliance with the regulations can be secured without it, a substantial saving in costs of administration secured, and increased dispatch attained in meeting the needs of the purchaser.

The grazing supervision work of the Forest Service is carried on under the same plan of organization and procedure as the timber activities, the aim being to bring about the best use of the forage crop from the standpoint of the public welfare. This necessitates a careful adjustment between range use and range productivity and also adjustment of range use to watershed, timber-production, and other requirements. " Grazing on the national forests is regulated with the object of using the grazing resources to the fullest extent possible consistent with the protection, development and use of other resources." [5]

As in the case of timber cutting and disposition, data are collected by the local staff in each forest and assistance is rendered by a force of grazing specialists moving from place to place. The data gathered with regard to grazing matters concern features such as land topography, quality and quantity of forage, predatory animals, facilities for watering, etc.

On the basis of the information obtained by this work (called range reconnaissance and range inspection) a system paralleling the timber plan is built up. The entire forest grazing area is divided into districts, just as the timbered area is divided into working circles. For these districts, management plans are worked out as in the case of the working circles, though with obvious points of difference.

Following this preliminary organization, comes the disposal and sale of this resource to the public. Forest rangers do not have the direct authority, as they do in the case of timber, to dispose of small portions of the resource, that power in the case of forage not descending below the grade of forest supervisor.

The Service activities in connection with the utilization of forage resources are confined to regulatory work. The stock is counted upon entering the forest. Thereafter the local forest officers see to it that the regulations are observed by the permit-holders during the time they are on the forest. Grazing permits (beginning as of January 1, 1925) generally run for ten years,

[5] Range management of the National Forests, Department of Agriculture Bulletin No. 790.

the government reserving the right to reduce the number of stock permitted thereunder in the interest of range protection or of the equitable treatment of other applicants. Over 7,700,000 cattle, sheep, horses, swine, and goats grazed on the forests in 1926, the property of about 28,000 permit holders.

As has been stated, forage utilization activities of the Service include the regulation of the use of the ranges by the permittees. In this regulation the Service is given assistance through the medium of livestock associations of permittees, upwards of seven hundred such associations being now in existence. Special rules pertaining to the management of livestock on the range are adopted by the associations, and approved and enforced by the Forest Service.

The present day utilization of the grazing resources is carried on under those clauses of the 1897 law providing for the regulation of the " occupancy and use " of the forests for all " proper and lawful purposes." Under the same clauses there has been developed a form of forest utilization known as " special use," in accordance with which small areas on the forests are rented from time to time to properly qualified individuals or associations for engaging in a wide variety of activities, both commercial and otherwise.

There are a number of enterprises on the forests conducted to the advantage of the owners as well as the forest communities they serve. They also return an income to the Service. Many of these enterprises are individual and unique and require special rates and conditions; while for others regular standards and fees have been developed.

The general test is that the purpose of the enterprise be " proper and lawful," under the language of the 1897 law, but legislation has also been enacted to cover the establishment of sanitariums and health resorts, as well as hotels, and summer homes.

Reasonably elaborate and careful plans for the development and supervision of this type of utilization are now made. Such plans, together with the supervision and inspection of the projects after they are established, to insure compliance with general forest regulations, and with the conditions incorporated in the permits granted, constitute the major portion of the direct work of the Service in this connection.

What has just been said applies also to the ordinary form of special use which requires the issuance of a permit and the collection of a rental or a fee. It applies only partially to that form of forest utilization known as " recreation," which, although it brings in little revenue, has come to be a forest use of the first magnitude. Automobile tourists passing through the forests create a problem in sanitation, fire hazard, and administration, the work of the local forest staff having been more than doubled in some instances as a result.

Approximately fifteen hundred camp grounds are located on national forest land. There is a difference between the forest summer camp and the forest tourist camp ground. The first is a tract of an acre or so allotted to an individual for a term of years under the 1915 act and improved at the allottee's expense with outing equipment. It brings in income to the government. The second is merely an area improved at the government's expense with simple toilets, piped water, etc., for the overnight or overweek stay of the motorist. This service brings no revenue but results in a definite expense. It does, however, minimize the danger to public health and property.

The last utilization activity of the Service is the supervision and control of the development, by private enterprise, of hydroelectric power projects on the national forests under the act of February 15, 1901; and giving assistance, in such supervision and control, to the Federal Power Commission which possesses the direct responsibility and performs part of the work. The first brings in some revenue through the Forest Service.

National forest utilization returned revenue amounting in 1929 to about $5,850,000 for timber and grazing, about 70 per cent of this total coming from timber. Special uses brought in about $340,000; water power permits about $100,000.

Most of the timber receipts come from commercial sales, cost sales producing only about $15,000. Turpentine operations produce about $15,000 annually; timber settlement cases a little less. Well over one billion feet of timber is cut from the national forests each year.

The bulk of the grazing fees comes from the grazing of cattle and sheep. About $15,000 comes from the grazing of horses, goats, and swine.

Protection. Some forty thousand forest fires occur in the United States each year. They burn over about ten million acres and cause an immediate annual property loss of about seventeen million dollars. In the national forests alone over seven thousand fires annually burn over four hundred and fifty thousand acres, and result in damage of over one and a half million of dollars. These figures are averages for the entire forested area of the country.

Insect damage to forest and shade trees in the United States amounts to $100,000,000 per year. Such damage to forest products increases this amount by $45,000,000. Certain tree diseases, such as the chestnut blight and the white pine blister rust, run the figures up an added $30,000,000. Of the present capital of saw timber in the United States it is estimated that over 300 billion board feet, or 14 per cent, must be discounted as cull timber due to decay induced by disease. Of the timber lost annually to the forests of the United States, public as well as private, it is estimated that $16\frac{1}{2}$ per cent is due to decay. Against all these menaces the forest must be protected, but fire constitutes probably the greatest problem of the Forest Service.

Fire. Fire protection in the forest is, basically, a problem of organization. The Service has a general fire policy for the entire territory under its charge. In conformity with that policy it works out a fire plan for every national forest and prepares supplemental plans from time to time for particular forest areas displaying unusual hazards or conditions. The data upon which these plans are made are collected by the local forest personnel.

A constant watch for fire is maintained during the dry season, especially in the regions of high hazard, where extra fire guards are employed. Arrangements are made for local aid in emergencies, and agreements are entered into with state forestry departments, timber protective associations of private owners, railroads, etc., for detection and suppression. Details of men from other Agricultural bureaus at work in or near the forests, such as the Bureau of Public Roads, are instructed to join the fire line at need.

In case of continuing extraordinary hazard, such as that represented by the great " Olympic Blowdown " in western Washington, such intensive measures as the patrolling of the dangerous area with auto trucks equipped with gasoline pumps and hose during the dry season, and the prohibition within the area of smoking or the use of fire in any form, are adopted.

In coöperation with the Weather Bureau the effort is made to keep all portions of the forests supplied with timely warnings when dry spells are impending or when air currents set in, which are known to be conducive to atmospheric conditions which heighten the fire hazard. Forest personnel is kept abreast of developing technique by conducting, during the off-seasons on the several districts, study courses for rangers in which forest fire protection is a major subject.

The larger contracts between the Forest Service and purchasers of national forest timber include a number of clauses to further fire prevention and suppression. Requirements upon the purchasers of large areas of timber call for patrol and safeguarding of the rights of way of railroads constructed to transport timber. The contracts also cover such matters as the restriction of smoking, or the building of camp or lunch fires on logging areas, or their total prohibition in times of danger, the strategic distribution of fire-fighting equipment, and the installation of such equipment upon locomotives, donkey engines, and other woods machines. In regions containing a considerable number of diseased or blighted trees the sanitary clause of the contract is so worded as to require the removal by the purchaser of all such trees within the area of his operations.

Considerable discretion is placed in the hands of local forest officers as to the rigidity of administration. Forest Service officers in charge of timber sales are required to be well trained in fire-control technique. The Service makes every reasonable effort to protect purchasers against loss of equipment, interruption of business, and the cost of suppressing fires for which they are not responsible.

The general fire code requires that a part of the fire plan for every national forest correlate the duties and responsibilities of Service employees and timber purchasers. For every large sale (and for other sales, when necessary) it is required that, as a supplement to the main fire plan for the forest, a subsidiary plan for the area under operation be drawn up by the local officers and the purchasers in conference.

Since 1919 the methods of fire-control just outlined have been supplemented irregularly by the use of the airplane, but the lookout station, the telephone, the patrolling ranger, and fire guard are still basic. So far the principal value of the airplane in American forest-fire fighting has been in connection with reconnaissance work. It has been found possible in the case of big fires to get information from various sectors of the fire in a half-hour's flying, which would not be obtainable in any other way.

Insects. Apparently no tree in America is insect proof, unless it be the sequoia of California. Trees in every forest region in the country are always threatened by insects, which, remaining quiescent for varying periods, attacking only sporadically, suddenly spread in immense infestations and destroy standing timber over wide areas.

The damage they do is both direct and indirect. They not only kill great stands of timber but also create areas of high inflammability in the forest.

In the combating of such infestations after they are discovered, recourse is had to the specialized knowledge of the Bureau

of Entomology. This Bureau, through its Division of Forest
Insects, is concerned with all large projects for the suppression
or control of serious infestations. The actual physical work
in such operations is performed, however, by Forest Service personnel detailed for that purpose, and there is a Service officer in
command or at least in close touch with operations. An entomolo-
gist from the Division is detailed to give advice as to the methods
of application. Methods of operation are illustrated by the following:

An important field of the activities of the Bureau of Entomology is the development of methods for the protection of our
forests from their insect enemies and the supervision of the application of these methods.

One example of the potentiality of forest-insect destructive-
ness is found in the work of the western pine beetle. During the
last 10 years, on an area of 1,165,000 acres in southern Oregon
and northern California, this beetle has killed 10 per cent of
the pine stand, causing a loss which amounts to 1,200,000,000
board feet valued at over $3,600,000.

In 1920 and 1921 this loss became so alarming that the private
timber owners and Federal Government combined forces to com-
bat the pest on their adjacent and intermingled holdings. For the
work on Federal lands and the entomological supervision Congress appropriated $150,000 and the private owners of timber
agreed to spend, if necessary, a like amount in the protection
of their timber.

May of 1922 witnessed the inauguration of this coöperative-
control project which without question is the largest of its kind
ever undertaken in the suppression of the western pine beetle.

The project area was divided into three parts and the actual
control work assigned to the United States Forest Service, the
United States Indian Service, and the Klamath Forest Protective Association (an association of the private timber owners), with general administrative supervision vested in a board
of control.

The remedial measures which had been developed by the
bureau many years ago were adopted for the project and the
Bureau of Entomology, Branch of Forest Insects, given authority to supervise the work to see that the methods were correctly
and properly applied. The methods consist, briefly, of locating

and felling the infested trees containing the live overwintering broods of the destructive bark beetles, peeling the upper part of the felled tree and burning all of the infested bark.[6]

Diseases. What has already been said about tree insect work gives a sufficiently accurate idea as to disease control method and detail in the Forest Service. The agents, however, are the Officers of Forest Pathology and Blister Rust Control of the Bureau of Plant Industry instead of the Division of Forest Insects of the Bureau of Entomology. The plant quarantine power of the Secretary of Agriculture is employed as occasion demands. Coöperation is maintained with interested states, with Canada, and with other bureaus of the national government, which have large forest areas under their control. Since 1909, when the Service first asked the Bureau of Plant Industry for assistance in connection with the appearance of the white pine blister rust, coöperation between the two bureaus along the same general lines obtaining in the Forestry-Entomology work has been constant. Consulting pathologists have been detailed to work in the national forests ever since.

Miscellaneous. Certain other activities of the Service are primarily of a protective nature, for example, game protection. Coöperation with state and local officers in the enforcement of the local fish and game laws was enjoined upon the Service in the first agricultural appropriation act passed after the amalgamation of 1905. Forest officers are frequently sworn in as local game wardens and thus clothed with state as well as federal powers. The protection afforded game animals, birds, and fish on the national forests is thus enhanced.

The work has also been furthered by the creation, by special enactment from time to time, of national game preserves situated wholly on government land and mostly within the boundaries of national forests. The Forest Service patrols and protects these areas, and looks after the animals preserved thereon,

[6] *Official Record,* September 12, 1923, p. 7.

such as the buffalo herds of the Pisgah and Wichita preserves, and the elk, deer, and antelope of the Wichita.[7]

In coöperation with the Bureau of American Ethnology of the Smithsonian Institution, the Service also furnishes protection from vandalism and thievery to the fifteen national monuments established within national forest boundaries under the Act for the Preservation of American Antiquities, of June 8, 1906.

These monuments have a total area of 378,145 acres, the largest being Mount Olympus, in Washington, which contains 298,730 acres, and is the natural habitat of the Olympic elk.

Another activity of a protective nature has to do with the combating of stock diseases, such as the " foot and mouth disease," which in recent years has been prevalent in the Southwest. An amendment to the rules and regulations governing the use of the forests, which was signed by the Secretary of Agriculture in August, 1924, empowered district foresters, in emergencies caused by this plague, to remove all stock from affected areas within their districts, and to close those areas to stock use indefinitely.

Under the various acts for the prevention of the unlawful removal or injury of national timber, the Forest Service also protects the forests from depredations, and under the general authority of law regarding the regulation of stream flow, especial protection is given to those forest watersheds which furnish water supplies for municipalities or afford the main supplies for irrigation or the development of power. The public health is also safeguarded by enforcing compliance with regulations forbidding the creation or continuation of insanitary conditions by persons using the national forests.

Development. The utilization and protection of a forest require the construction or installation therein, of certain physical improvements, the eradication or removal of undesirable or hazardous elements, and the modification of existing natural

[7] Department of Agriculture, The Wichita National Forest and Game Preserve. Miscellaneous Circular No. 36.

7

features into forms consonant with the scheme of the enterprise. From the beginning the Forest Service has developed the property entrusted to its care in order "to improve and protect the forest," to secure "favorable conditions of water flows, and to furnish a continuous supply of timber for the use and necessities of citizens of the United States."

Roads and Trails. Types of roads in the national forests vary according to purpose or use. There are those designed primarily for the convenience and the development of local communities, surrounded or partly surrounded by the forests, those which are included in nation-wide highway systems which incidentally cross the forests, and those designed primarily to make the forest accessible for protection, recreation, and economic use. However, in the Rules and Regulations of the Secretary of Agriculture for Administering Forest Roads and Trails under the Provisions of the Federal Highway Act of November 9, 1921, these roads are classified as forest highways and forest-development roads. Each kind of road under this classification is further subdivided as to major projects and minor projects. By far the greatest number of forest-development roads are minor projects; but certain forest-development roads of the more difficult sort (specifically those costing more than five thousand dollars per mile as an average) are classified as major projects. Also, certain of the forest highways of a more simple nature (specifically those not requiring the technical services of a highway engineering organization or costing more than two thousand dollars per mile) are classified as minor projects.

Practically all the road work of the Forest Service is concerned with these minor projects, which include trails. With the major forest-highway projects the Forest Service is concerned when appeal for aid in forest-highway construction has been made by a local government or a local association or interest of any kind. In such cases the Service considers the proposed project from the standpoint of public need, but particularly from the forest standpoint, that is, the desirability of the proposal as respects forest utilization, protection, and administration, and makes recommendation to the Secretary of Agriculture. Decision as to the undertaking of the project lies with the Secretary, who draws his conclusions first from the report of the state highway commission and then from the reports of the Forest Service and the Bureau of Public Roads. Once the construction of a major project is undertaken the Forest Service has no further concern with it, except in such matters as clearing and brush disposal. Construction and maintenance fall within the jurisdiction of the Bureau of Public Roads.

Most minor projects are selected, surveyed, constructed, and maintained by the Forest Service alone. In only two respects does another agency have a relation to that work. Where a minor project consists of one of the lesser forest highways, the Bureau of Public Roads has a voice in its selection. Again, in the case of a minor highway project that is likely to be enlarged and improved later on so as to become part of an important highway system, the survey work is performed by the same bureau. In constructing these minor projects the Service utilizes the local forest organizations in conjunction with its regular administrative, protective, and other activities. The work is coördinated, so far as possible with fire protection, so that road construction crews may be available in regions of high hazard as part of the fire suppression organization. Under a coöperative agreement employees of the Bureau of Public Roads may be utilized for fire-fighting if emergency situations develop in regions where they are engaged.

In selecting projects for the outlay of the forest development fund, the Forest Service gives consideration to the following factors: comparative existing transportation facilities, comparative value of timber or other resources to be served, comparative difficulties of road and trail construction, and relative fire danger. In this connection the Forest Service, for a number of years, has been engaged in a study of road requirements on the several forests. From the data so secured, together with the recommendations from the state and counties affected, the program for road development is progressively built up.

The Forest Service is free to call on the Bureau of Public Roads for technical advice in meeting difficult problems in forest-development road construction, and it has the privilege of renting equipment from the stocks of that Bureau. The Bureau does not permanently maintain any employees on the national forests for that part of the forest road work which it handles. However, in the district offices of those districts of the Bureau which contain national forests, there is a permanent staff of engineers, superintendents, and clerks assigned wholly to forest work.

Other Construction. Besides road and trail improvements the Forest Service constructs fire breaks at hazardous points. The Service also constructs, partly with its own resources but generally on a basis of equal shares with livestock associations, under coöperative agreements, a considerable mileage of stock driveways for the passing of stock from one part of a forest to another or across a forest.

There have also been constructed, partly by the Service directly, partly by private persons and companies assisted by the administrative use of timber, about 37,000 miles of telephone line, which are of assistance for administrative purposes and protection against fire.

Other improvements to the forests include such projects as lookout towers; ranger station buildings for the accommodation of forest rangers and their families, stock, and machinery; equipment stations; camps for the accommodation of tourists; rough bridges for the trails and simpler roads; stock fences; stock water developments; stock corrals; and the necessary buildings for forest nurseries. The maintenance of these improvements, as well as their original construction involves the Service in continuous activities.

The improvements having to do with stock may be constructed in coöperation with livestock associations, the members of which have grazing privileges on the forests. In a few instances also, tourist camps are constructed in coöperation with municipalities, commercial clubs, etc. Within recent years the Service has responded to the growing demands of the tourist by the retention, from time to time, of recreation engineers for the better planning of the public camp grounds.

Miscellaneous Work. The forest may be developed by the eradication of undesirable features as well as by the construction of desirable ones. Work along these lines includes the eradication of gooseberry and currant plants (the intermediate hosts of the blister rust) and of certain forage plants and grasses poisonous to stock. This forage work has the technical coöperation of the Bureau of Animal Industry. Much of the actual removal is done by stock permit-holders under the coöperative agreements.

Considerable work in connection with the wild life on the forests is performed by the Forest Service in coöperation with the Biological Survey, in the extermination of predatory animals, using bait, traps, or ammunition furnished by the Survey, and in poisoning harmful rodents.

The Forest Service plants about twenty thousand acres each year, principally with seedlings grown in Service nurseries, but to some extent by direct seeding. Something over one hundred thousand acres of forest land have been successfully reforested by planting. The nursery phase of this form of forestal development has varied considerably. Since 1911 the tendency has been to concentrate the work in larger nurseries, with one nursery to each section of the country with a distinctive forest cover.

The timber utilization contracts made by the Forest Service constitute an important instrument for the furthering of the forest policy of the nation in three major aspects. They permit utilization of the stock of ripe timber, they afford assistance in fire, insect, and disease protection, and they promote forest development.

Under the Forest Homestead Act of June 11, 1906 (34 Stat. L., 233), the Service, with the occasional technical assistance of the Bureau of Chemistry and Soils, examines lands within the boundaries of the forests to ascertain whether they possess sufficient agricultural value to warrant throwing them open to homesteading.

Likewise, under the act of March 20, 1922 (42 Stat. L., 465), the Forest Service performs the necessary preliminary work in the way of appraising, surveying, and cruising called for in the land and timber exchanges authorized under that act for the purpose of consolidating the national forests.

The forest aspects of the acquisition of lands by purchase, for inclusion in the national forest system, under Section 6 of the Weeks Act as amended by the same section of the Clarke-McNary Act, are attended to by the Service. The duties included are as follows:

. . . the Secretary of Agriculture is hereby authorized and directed to examine, locate, and recommend for purchase such lands as in his judgment may be necessary to the regulation of the flow of navigable streams, and to report to the National Forest Reservation Commission the results of such examination: *Provided,* That before any lands are purchased by the National Forest Reservation Commission said land shall be examined by the Geological Survey and a report made to the Secretary of Agriculture, showing that the control of such lands will promote or protect the navigation of streams on whose watersheds they lie.

The legal aspects of the transactions are also looked after by the Forest Service in a sense, since the law personnel which does the work, though not on the Forest Service payroll (being actually part of the force of the Solicitor for the Department of Agriculture) is attached permanently to the Service for performing the title work involved in the acquisition of the purchase areas.

The procedure involved in the acquisition of these lands is as follows: The Geological Survey examines a region to determine whether it comes within the purview of the navigable-waters feature of Section 6. The Forest Service then examines and appraises particular tracts offered for sale in that region. On the basis of the field examinations, and upon a tentative agreement as to terms, the Secretary of Agriculture recommends for purchase. The National Forest Reservation Commission then approves for purchase, and the purchase contract is executed and survey and title examinations are completed, the first by the Forest Service, the second by the legal staff mentioned above. The Attorney General approves the title. Conveyance of title and payment are then consummated.[8]

Investigation. The research activities of the Forest Service may be discussed under three general heads: those having to do with the grazing or range-management phase of the work of the Service; those having to do with timber and timber products; and; those which have to do with the economic aspects of the forest problem. The economic studies are of recent origin; the others are old, established projects.

Range Research. Range research includes the collection of range-forage data and the development of practical and economic range-management practices. The handling of livestock under varying conditions, the development of systems of grazing periods, watering, salting, seasonal readiness of plants for forage, eradication of poisonous range plants, reseeding of the range, range carrying capacity, and the harmonizing of grazing with watershed protection, timber production, wild life, and other resources of forest and range land—these problems and many others are studied by the Forest Service grazing research personnel, both independently and in coöperation, particularly with the Bureaus of Plant and Animal Industry.

Timber and Timber Products. In the forest regions forest research is carried on through a number of experiment stations which perform the same function in relation to timber growing that the agricultural experiment stations do for farm-crop raising. The stations study and determine the technical and scien-

---

[8] See Department of Agriculture booklet on "Purchase of land for national forests under the act of March 1, 1911, The Weeks Law," as revised May 1, 1921, pp. 4-5.

tific bases for the development and protection of the forest. Investigations on artificial forestation for various species are made. Rate of growth is studied and the requirements of species as to soil, sunlight, thickness of stand, etc., for proper growth, as well as cutting and thinning methods for the different forest regions. Desirable kinds of trees for growth under prevailing conditions and how such trees may be most easily and economically grown are other problems. Efforts are made to improve nursery practice in the growing of seedlings for planting in devastated areas and in hand-planting. The object is to improve planting practice, increase knowledge of artificial forestation from selection of seed or species to methods of seeding or planting, and reduce costs. The protection of the forests from fire, insects, and diseases, is under constant investigation.'

The greater part of the research effort of the Service with respect to timber and its products is carried on at the main Service laboratory at Madison, Wisconsin, though considerable field work is also performed in different parts of the country. This research work covers a wide range, and is mentioned here only in so far as it involves certain sharply defined features, such as the chemical, physical, and pathological aspects of wood.

In timber mechanics, for example, the strength of wood and wooden articles, together with their toughness and elasticity, are studied. Tests are made of clear specimens, of knotty, checky, and shaky specimens, and of specimens subjected to practically every variety of preservative treatment. Tests are also made of barrels, boxes and other containers, both filled and unfilled, and vehicle, implement, and airplane parts. In this way data are worked out and made available for manufacturers touching the fitness for use of various substitute species, and the possibilities for increased strength and durability resulting from improvements in design.

Other tests are made of density, shrinkage, transfusion of moisture, specific heat, heat conductivity, heat of absorption of water in wood, and the permeability of wood by liquids and gases. Collectively these forms of wood-testing are known as experiments in timber physics, which also include the identification of woods, the Laboratory being called upon to identify several thousand specimens each year.

In wood-preservation work studies are made of the several preservatives now in use, such as creosote and zinc chloride, and of the various processes utilized in the application thereof; with special attention to the various factors entering into the processes, such as temperature and pressure. Studies are also made in this connection of the fire-proofing of wood, and of the preparation and application of varying glues for use in the process of lamination.

One of the most important fields of investigation at the Laboratory is the use of wood in the manufacture of pulp for paper-making. Processes for pulp grinding and cooking are studied, as are the potentialities for pulp-making in certain unused species, or in the limbs, bark, sawdust, and slabs now classified as waste after lumbering and milling. The pulping possibilities of materials other than wood, such as flaxstraw, cotton linters, and oat hulls, are also studied; and experiments are carried on to determine the utility of wood pulp for making buttons, lacquered articles, electrical fittings, smokeless powder, artificial silk, etc.

The chemical aspects of wood offer a similarly wide field, the possibilities of obtaining wood derivatives by distillation and other forms of conversion being observed at the Laboratory. In addition, researches into the chemical composition of wood itself are conducted.

Experiments in waste elimination by chemical, physical, and mechanical means are supplemented in the forests and at lumber plants by organized efforts for more efficient utilization in the trees and in boards, and timbers. One of these is the matter of "dimension stock." The Laboratory carries on investigations as to the possibilities of cutting, direct from the log, the exact sizes to be used in manufacture, thus eliminating part of the waste involved in cutting from standard lumber sizes at the factory. Likewise, investigations into the use of the small sizes and low grades of lumber are being conducted.

Finally, there are being carried on at the Laboratory, by a detail from the Bureau of Plant Industry, investigations into the durability and decay of wood, the effect of preservatives, the reasons for the development and growth of fungous excrescences, the rotting of building timber, the sanitary storage of lumber, etc.

Economic Research. In addition to grazing, forest management, and forest products, the research program includes many other subjects. These include forest-taxation systems, insurance against fire, the costs and returns in private forest practice, surveys of states and industrial districts with respect to the timber and lumber situation, studies of financing, marketing, and distributing lumber, and the collection of general statistics concerning lumber and allied products.

One of the most important and complex projects ever undertaken by the Forest Service is the forest survey just being initiated. This is a nation-wide, comprehensive determination of forest resources; of present and potential growth; of the drain upon the forests not only by cutting but through fire, disease, and insects; and of the present and future needs of the country for forest products.

Most of this work is on a coöperative basis. Certain surveys are conducted independently, others with state and private coöperation. Some statistical work is performed in coöperation with the Census Bureau. The forest survey involves coöperation with many public and private agencies.

Miscellaneous. Many Forest Service activities, by reason of connection with a coöperating organization or agency, may be designated as "research," though the Service merely assists. Examples occur in connection with the aid which the Forest Service gives to the Biological Survey in collecting wild-life data to be used in Survey calculations and investigations. Similar work is performed for the Bureau of Fisheries. The Forest Service also coöperates in scientific fields, upon request, with other organizations and agencies.

Coöperation. Mention has been made of the contact and coöperation, apart from research, which the Service maintains with other governmental agencies in the furtherance of the work of both.

In addition to interdepartmental coöperation the Forest Service maintains contact with bureaus of other departments, particularly in inspectional work or work involving some phase of specialized knowledge. The Service inspects wooden parts for airplanes, the seasoning of wood bought for governmental use, and other forest-product stores for the same purpose. It assists the Bureau of Reclamation in forest phases of reclamation work, does similar work for the Office of Indian Affairs in connection with land allotments, and makes examinations of timber-purchase areas on Indian reservations. It also coöperates, to some extent, with the Indian Service in regard to fire-fighting in Indian timberlands, and considerable assistance of this nature is given the National Park Service. It works with the Department of Commerce in the standardization of lumber grades and names and with the Bureau of Fisheries in stocking forest streams with fry, and has contact, in regard to law enforcement, with the Department of Justice.

Perhaps the most important example of Service coöperation is the work carried on with various states in fire prevention and suppression. Coöperation, in this work, is maintained with thirty-eight states, containing, within their borders, a total area in need of protection of 380,000,000 acres. The Service also coöperates with a number of states in the study of taxation problems concerning forests, as provided for in Section 3 of the Clarke-McNary Act. Under Section 4 of the same act (providing for coöperation with states for the procurement, production, and distribution, to farmers, of forest-tree seeds and plants for the purpose of establishing wind breaks, shelter belts, and farm wood lots upon denuded or non-forested lands) the Forest Service is working with thirty-nine states. The work involved in the use of such material after distribution (as provided for in Section 5 of the Clarke-McNary Act), that is, farm forestry proper, has been placed within the province of the Extension Service of the Department of Agriculture.

Finally the Service coöperates with states and private organizations in ways not covered by particular enactments. In New Mexico, for example, under the terms of a special agreement between the Department of Agriculture and the New Mexico Commissioner of Lands, the Service is administering the timber utilization on 215,000 acres of state-owned land, containing some 550,000 feet of timber, according to the practice which has been evolved for the national forests, and has taken over in large measure the forest work of the state. In the same state the Service has coöperated with the New Mexico Archaeological Society in restoring the Bandelier National Monument.

Information. The Forest Service diffuses information about its work and endeavors in numerous ways to better forestry practice in general. Through its publications it seeks to place before the public the fruits of its experience, study, and research. By giving advice to woodland owners (when solicited), and to manufacturers, dealers, and users of forest products it seeks to widen its field of usefulness.

In addition to the regular Service publications, contributions are made by Service personnel to forest-school, trade, and other publications. Articles are published in the newspapers and newspapermen are supplied with data.

Exhibitions are made at state and county fairs throughout the country illustrative of Service activities and resources. These exhibits, which cover practically the entire range of the work of the Service, are especially directed at two objectives: wood-lot development and management, and prevention of forest fires.

The forest-fire propaganda is disseminated in a variety of ways; by publications, by the distribution, lending, and exhibiting of photographs, lantern slides, and motion pictures, and by coöperation with interested private organizations and other bureaus and departments of the national government which are in a position to assist. The work is furthered, likewise, by the encouragement of the formation of forestry clubs of boys and girls, particularly in the heavily timbered regions, by both the Forest Service and the Extension Service.

An effective phase of forest-information dissemination includes the courses for manufacturers and dealers in forest products, which are given at the Madison Laboratory. These courses include such subjects as gluing, kiln-drying, boxing and crating, wood properties, wood as a building material, and box wiring and metal binding. There are also courses for lumber salesmen, designed to give general technical information on the properties of wood and specific data on the species in which the attending students are especially interested, to the end of utilizing lumber for the purpose to which it is best suited and eliminating inefficient utilization as a source of waste.

Acquisition. Although the acquisition of new forest lands is not solely a Forest Service function, the Service participates in such acquisition as the field agency of the National Forest Reservation Commission, and is directly affected by such acquisition.

CHAPTER III

ORGANIZATION [1]

The Forest Service is a bureau of the Department of Agriculture headed by the Forester, who is directly responsible to the Secretary of Agriculture. The Service is organized partly on a functional and partly on a geographical basis, neither phase of organization being exclusive.

Broadly speaking the primary work of general administration is carried on at Washington while the detailed administration, the bulk of experimental work, and the technical operations inhere to the field.

General Administration. The executive and administrative work at Washington, other than that directly concerned with technical operations, is conducted through the Office of the Forester, which may be said to include (1) The Forester, (2) the Associate Forester, and (3) the Branch of Finance and Accounts. This Office, through the District Foresters, also maintains executive supervision over the Field Administration.

Office, Proper, of the Forester. The Office, proper, of the Forester consists of the Forester, the Associate Forester, and the necessary secretarial and stenographic service. The Associate Forester acts as general executive assistant, relieving the Forester of detail and acting, in his absence, as chief of the Service. The Associate Forester also makes field inspections and substitutes for the Forester in many matters, including public appearances, hearings, etc.

The Forester serves, ex-officio, as a member of the Forest Protection Board and the National Capital Park and Planning Commission.

[1] This chapter shows the organization as it was in April, 1930. It, therefore, differs in some minor details from Appendix 1, which is as of November, 1929.

Branch of Finance and Accounts. The Branch of Finance and Accounts is headed by a Chief and an Assistant Chief, and is concerned with accounting, auditing, cost keeping, and receipts and disbursements. It also arranges for personnel transportation and conducts the work involved in Service appointments. Two accountants, an auditor, a transportation and rate clerk, and an appointment clerk are employed in addition to the necessary clerical assistants.

The various subdivisions of the work may be indicated as follows:

Accounting: General	Audit
Accounting: Roads	Claims
Accounting: Acquisition	Cost Keeping
Appointments	Transportation

The direct administration of technical operations is carried on through seven coördinate major units known as branches. These are:

1. Branch of Operation
2. Branch of Forest Management
3. Branch of Range Management
4. Branch of Lands
5. Branch of Engineering
6. Branch of Public Relations
7. Branch of Research

Each branch is under direction of an Assistant Forester, except Engineering, which is headed by a Chief Engineer. These officers are directly responsible to the Forester.

Branch of Operation. The Branch of Operation is charged with Servicewide supervisory and inspectional duties. It conducts administrative studies looking toward improvement of methods and institutes inspections to test administration. It supervises finance and accounts from an administrative standpoint, has general charge of personnel, directs the work of fire control for the Service at large, purchases stores, and distributes supplies.

This Branch has charge of all improvement work in the forests except road building. It is responsible for the construction and maintenance of ranger cabins, lookout houses, experiment stations and nursery buildings, telephone lines, fire-fighting stations and equipment, tourist camps, and other accessories essential to the work of forest protection and forest utilization.

At the head is an Assistant Forester, who is aided by an Assistant Chief. The general Administrative force includes an Inspector, an administrative assistant, and several clerks.

The technical work is allocated to two major subdivisions: (1) The Office of Maintenance and (2) the Supply Depot.

Office of Maintenance. The Office of Maintenance is headed by the Chief of Maintenance and the personnel of the main force includes such employees as purchasing agent, mail clerk, electrician, and carpenter, in addition to the usual clerical and telephone service.

This Office also includes two sub-divisions, or sections: (1) The Section of Stenography and Typing, and (2) the Section of Supplies.

Supply Depot. The Supply Depot is located at Ogden, Utah. It is headed by a Supply Officer, who is aided by such employees as a printer, clerks and stenographers, an instrument-repairman, a mechanic, and packers and laborers. This Depot, in the geographical center of the Western forest area, maintains a stock of all supplies needed in the several forest districts, which are distributed to the districts upon requisition.

Branch of Forest Management. The Branch of Forest Management is concerned with the formulation of plans and policies and the supervision of methods under which the timber resources of the national forests are utilized and the protective function of timber growth as watershed cover is exercised. The Branch performs staff work in connection with the establishment of working circles and the development of cutting plans. It later exercises general supervision over the execution of such policies and plans and, in effect, controls the sale and cutting of national forest timber. This Branch also supervises the reforestation of denuded land, in connection with which it maintains nurseries in the national forests to collect seed and grow planting stock. The more important of these nurseries are located as follows:

1. Savenac—Haugen, Montana
2. Bessey—Halsey, Nebraska
3. Cass Lake—Cass Lake, Minnesota
4. Beal—East Tawas, Minnesota
5. Monument—Monument, Colorado
6. Wind River—Stabler, Washington
7. Parsons—Parsons, West Virginia
8. Susanville—Susanville, California

An Assistant Forester heads the Branch of Forest Management, and he is aided by an Assistant Chief, and a Forest Inspector. The necessary complement of clerks is employed.

Branch of Range Management. The Branch of Range Management is headed by an Assistant Forester, who is assisted by two Inspectors and a clerk.

This Branch conducts the work in connection with the administration of the forage resources in the forests, including the letting of grazing privileges and the division of ranges as to various owners and classes of stock. Certain phases of fish and game work are carried on, both independently and in coöperation with local governments, the Bureau of Biological Survey, and the Bureau of Fisheries. The Branch also supervises the reseeding of depleted ranges or their restoration by alternating grazing periods, and, in coöperation with federal or state officers, the regulation of quarantine on livestock.

Branch of Lands. The primary function of the Branch of Lands is to build up the national forest properties (1) by additions and eliminations, (2) by the acquisition of privately owned lands through exchange or purchase, and (3) by appropriate action against claims which do not conform to law; in other words, the betterment of forest areas by appropriate extensions or contractions of boundaries and by preventing illegal appropriations, on the one hand, and by determining, on the other, what national forest lands may properly pass to private

8

ownership. The Branch is responsible for the leasing of forest areas on term and special use permits under the general provisions of the 1897 law, and of the acts of February 28, 1899 (30 Stat. L., 908), and March 4, 1915 (38 Stat. L., 1095, 1101). It supervises the work necessary to the selection, classification, and segregation of lands within the national forests which may be opened to settlement and entry under homestead laws applicable to the forests. It is concerned with questions involving claims to tracts of land within the forest boundaries under the mining, homestead, and other statutes, and the exchange of national forest lands or timber for lands in private ownership within the forest boundaries. It also handles all administrative features in connection with the acquisition of forest lands by purchase under the provisions of the Weeks and Clarke-McNary laws.

The Branch is headed by an Assistant Forester, aided by an Assistant Chief. Land law examining and other clerks are also employed.

Branch of Engineering. The Branch of Engineering surveys and maps the national forests for the purposes of utilization, protection, and development; coöperates with the Geological Survey, the Coast and Geodetic Survey, and other organizations engaged in general surveying of national forest lands; locates and supervises the construction of such roads as are handled by the Forest Service; administers for the Service the pertinent provisions of the national forest road appropriation acts and regulations; makes, for the Federal Power Commission, such investigations and reports concerning projected and licensed power developments on the national forests as are called for under the terms of the act of June 10, 1920 (41 Stat. L., 1063); and administers power permits and easements under the terms of the acts of February 15, 1901 (31 Stat. L., 790), February 1, 1905 (33 Stat. L., 628), and March 4, 1911 (36 Stat. L., 1235, 1253).

All civil engineering and topographic and cadastral surveys are directed by the Branch of Engineering, which also serves the entire Forest Service through the maintenance of the Forest

Atlas, the compilation of Forest Service maps, and the performance of all required drafting and photographic work.

The Branch is under the direction of a Chief Engineer, who is aided by an Assistant Chief Engineer and by an Assistant Engineer. The technical work is carried on through three subordinate units:

        1. Office of Maps and Surveys
        2. Office of Roads
        3. Office of Water Power

Office of Maps and Surveys. The Office of Maps and Surveys is headed by the Assistant Chief Engineer, who is aided by an assistant. The Office is composed of five sections:

        1. Drafting
        2. Atlas
        3. Lithographing
        4. Photography
        5. Surveys

The Section of Drafting is headed by a Chief Draftsman and employs a force of draftsmen, an artist, and a file clerk.

The Section of Atlas is headed by a Statistician, who is assisted by a draftsman and a clerk on the compilation of a National Forest Atlas. This atlas consists of many loose-leaf volumes containing information, in elaborate detail, regarding the national forests. These data have been collected since the origin of the Forest Service and are kept up-to-date. They include complete maps of the forest regions, all proclamations and executive orders affecting them, all information available on timber forage, and other resources, annual cost and income, improvements, progress of reforestation, natural and artificial, and other pertinent matters of varied nature.

The Section of Lithographing is headed by the Assistant Chief of Maps and Surveys, who supervises the reproduction of Forest Service maps by lithography.

The Section of Photography is headed by a Chief Photographer, aided by an administrative clerk and a number of photographers.

The Section of Surveys is headed by an Assistant Engineer.

Offices of Roads and of Water Power. The Offices of Roads and of Water Power are headed, respectively, by the Assistant Chief Engineer and by the Chief Engineer.

Branch of Public Relations. The Branch of Public Relations was created " to promote the diffusion of knowledge of forestry generally, as well as to secure the best results in the efforts of all parts of the Forest Service to bring about better protection of the national forests." It is responsible for devising and developing the means of contact with the public, giving out information to the press, preparing, printing, and publishing the results of the Forest Service's varied activities, and showing and distributing forestry exhibits, pictures, lantern slides, and moving picture films.

This Branch coöperates with the states and with private enterprises in connection with the protection of forest lands from fire under the provisions of the Weeks and Clarke-McNary laws; that is, in planning, developing policies, and improving protective measures. It also supervises the work of the Forest Service in improving forest management under the owners of farm wood lots and maintains a current compilation of state forest statutes. It appraises state policies and methods of control as a basis for determining the apportionment of federal aid under the acts above mentioned.

The general administrative direction of this Branch is under an Assistant Forester and Chief of Branch, who is aided by an Assistant Forester for Special Assignments and a Logging Engineer for Coöperation with Timberland Owners.

The general technical work of the Branch is allocated to two Divisions: (1) State Coöperation, and (2) Information.

Division of State Coöperation. The Division of State Coöperation is headed by a Forest Inspector and Chief of Division, aided by a second Forest Inspector, an Assistant Inspector, a Law Compiler, and several clerks. This staff handles the general administrative work of the Division. The detailed work is divided between two units designated as (1) Fire Protection and Distribution of Planting Stock, and (2) Farm Forestry.

The Fire Protection and Distribution of Planting Stock unit carries on its work under a geographical organization divided as to:

1. Northeastern States—Amherst
2. Middle Atlantic States—Washington, D. C.
3. Southeastern States—Asheville
4. Gulf States—New Orleans
5. Central States—Louisville

Each district enlists the services of a District Inspector.

The Farm Forestry unit is headed by an Extension Forester. It is conducted in coöperation with the Office of Coöperative Extension Work.

Division of Information. The Division of Information of the Branch of Public Relations is under the direction of a Chief of Division, who is aided by an Editor, for press relations, and an Artist. The technical work is divided among three subdivisions as follows: (1) Office of Educational Coöperation, (2) Office of Publication, and (3) Office of Visual Education.

The Office of Educational Coöperation is supervised by an Editor, who is aided by editorial clerks and correspondence clerks. The Office of Publication is headed by a Chief of Publication. It also includes a Section of Printing. The Office of Visual Education is directed by a Chief.

Branch of Research. The Branch of Research is concerned with the administration and technical aspects of all research activities of the Forest Service, including forest experiments, forest economics, forest products, industrial investigations, dendrology, range investigations, forest measurements, and forest statistics.

It is headed by an Assistant Forester, aided by an Assistant Chief of Branch and an Editor.

This Branch is organized on a functional basis, the technical work being carried on through the agency of five subdivisions:

1. Office of Silvics
2. Office of Forest Economics
3. Office of Forest Products
4. Forest Products Laboratory
5. Office of Range Research

Office of Silvics. The Office of Silvics is organized partly on a functional and partly on a geographical basis, with one group of units which might be called the home office and another the field. At the head is a Chief, aided by two technical assistants. Also located at Washington are two sections concerned with general service or supervisory work: (1) Section of Forest Measurements, and (2) Section of Library.

The Section of Forest Measurements lists a Chief of Section in charge of forest mensuration and statistical work, two Assistant Silviculturists, and a Junior Forester. The remaining personnel consists of statistical clerks and tabulating machine operators. The Section of Library consists solely of the Librarian of the Office.

The field division of the Office is composed of eleven stations in which forest experiment work is carried on:

1. Allegheny
    Territory: New Jersey, Pennsylvania, Maryland, and Delaware.
    Headquarters: Philadelphia.
2. Appalachian
    Territory: Virginia, West Virginia, eastern Kentucky and Tennessee, North Carolina, and South Carolina.
    Headquarters: Asheville.
3. California
    Territory: California and western Nevada.
    Headquarters: Berkeley.
4. Central States
    Territory: Ohio, Indiana, Illinois, Iowa, Missouri, and western Kentucky and Tennessee.
    Headquarters: Columbus.

5. Lake States
    Territory: Michigan, Wisconsin, Minnesota, North Dakota, and South Dakota.
    Headquarters: St. Paul.
6. Northeastern
    Territory: New England and New York.
    Headquarters: Amherst.
7. Northern Rocky Mountain
    Territory: Montana, and northern Idaho.
    Headquarters: Missoula.
8. Pacific Northwest
    Territory: Washington and Oregon.
    Headquarters: Portland.
9. Rocky Mountain
    Territory: Nebraska, Kansas, Wyoming, and Colorado.
    Headquarters: Colorado Springs.
10. Southern
    Territory: Georgia, Florida, Alabama, Mississippi, Arkansas, Louisiana, Texas, and Oklahoma.
    Headquarters: New Orleans.
    Substation: Starke, Florida.
11. Southwestern
    Territory: New Mexico and Arizona.
    Headquarters: Flagstaff.

In general these stations are similarly organized, each being headed by a Director, usually a trained forester, who is aided by investigative assistants and clerks.

Office of Forest Economics. The Office of Forest Economics, under a Senior Economist, is organized with a general economic and administrative staff and a number of special project groups. The general group includes three Forest Economists, a Statistician in Forest Products, a statistical clerk, and stenographic and clerical aid.

The Forest Taxation Inquiry Unit maintains headquarters at New Haven, Connecticut. The group is directed by an Economist, under whom are investigative assistants, mathematicians, and clerks. It is concerned with tax studies, state lumber surveys, prices, etc.

The Forest Survey, included in the field of forest economics, is in immediate charge of a Director. The field work is handled by regional directors who, together with technical staffs, have been added to the forest experiment stations.

The project entitled Financial Aspects of Private Forestry is carried on by a staff of four technical men with clerical assistants under the direct charge of a Senior Forest Economist, assigned to the staff of the Southern Forest Experiment Station.

The Insurance study is handled by a Senior Forest Economist and one technical assistant assigned to the staff of the Pacific Northwest Forest Experiment Station.

Office of Forest Products. The Office of Forest Products is concerned with the broad and general aspects of lumber and wood products and their uses, as well as experimentation therein and extension thereof. It is headed by the Senior Engineer in Forest Products, who is aided by a scientific assistant, an associate statistician, and the necessary stenographic and clerical help.

Forest Products Laboratory. The Forest Products Laboratory is located at Madison, and is conducted in coöperation with the University of Wisconsin. The subdivisions of this laboratory organization may be classified under two general heads: (1) Those concerned with administrative and general service duties, and (2) those conducting the primary or technical work:

General Service

1. Office of the Director
2. Section of Laboratory Operation
3. Section of Finance and Accounts
4. Section of Publication of Results.

Technical

1. Office of Dry Kiln Expert
2. Section of Pathology
3. Section of Silvicultural Relations
4. Section of Derived Products
5. Section of Pulp and Paper
6. Section of Timber Mechanics
7. Section of Timber Physics
8. Section of Wood Preservation
9. Section of Industrial Investigations

The Office of the Director includes the Director, who is executive head of the Laboratory, a principal chemist, a principal engineer, and a secretarial clerk.

Through this Office is conducted the work of general administration and coördination both as to technical and other operating activities. As immediate aid to this Office and as a general service unit for the entire Laboratory the Section of Laboratory Operation has been established. This Section handles records, supplies, personnel detail, computations, engineering matters, photography, and quarters maintenance, and exercises a general administrative supervision of finances.

The Section is headed by a Chief of Section, an engineer, but the organization lists no definite subdivisions or sections. The employees are numerous and their duties are varied.

The Section of Finance and Accounts, second of the " general service " units, deals with the matters indicated by its title. This Section is directed by a Fiscal Agent and Section Chief, who is aided by a deputy fiscal agent. There are also employed a time and cost-property clerk and a stenographic and general clerk.

The Section of Publication of Results attends to reviewing and editing duties, press service, multigraphing, giving out information, collecting and preserving technical notes, making illustrations, and distributing Laboratory publications; that is, the general dissemination of the results of all Laboratory research work. It is headed by a Senior Forester, who is associated with two senior technical reviewers, a technical reviewer, and a forester. In addition, the Office employs a senior scientific illustrator, a multigraph foreman, multigraph operators, and stenographic and typing clerks.

The technical divisions of the Forest Products Laboratory are nine in number. One has the title of " office " while the others are called " sections."

The Office of Dry Kiln Expert consists of one employee, a Physicist and Expert in Kiln Drying, his work being that of experimentation in this method of seasoning timber and timber products.

The Section of Pathology is conducted in coöperation with the Bureau of Plant Industry, and consists of one employee, a stenographic clerk.

The Section of Silvicultural Relations conducts studies to determine how the properties of wood are affected by growth conditions and by structure, and to determine how growth conditions affect resin production. This section is also concerned with wood identification. It is headed by a Senior Specialist in Wood Structure, aided by a senior microscopist in forest products, a silviculturist, an assistant physiological plant anatomist, an assistant wood technologist, a senior xylotomist, and an assistant laboratory aid and translator.

The Section of Derived Products is headed by a Chief of Section, a senior chemist, who has associated with him, as technical aids, two senior chemists, two assistant chemists, and four junior chemists. An under laboratory aid and a stenographic and general clerk are also employed.

The subsections of this Section may be designated as follows:

1. Chemistry of Wood Preservatives
2. Hydrolysis and Wood Extractives
3. Physical Chemistry of Wood
4. Micro-chemistry of Wood
5. Movement of Liquids in Wood

The Section of Pulp and Paper conducts investigations in the field indicated by the title, including plant growth other than wood. The Section is directed by a Chief of Section, a principal chemist, who is aided by an Assistant Section Chief, a chemist.

The Section of Timber Mechanics, concerned with experiments in the strength, toughness, and elasticity of wood and wood products, is headed by a Chief of Section, who is a principal engineer. The staff includes also engineers of several grades, laboratory aids, assistant laboratory aids, stenographic clerks, and a mechanic.

The Section of Timber Physics conducts work on kiln drying and air drying. A Chief of Section, who is a principal engineer, directs the work, with an Assistant Section Chief. There are also engineers of several grades, an associate wood technologist, laboratory aids, and a stenographic and general clerk.

The Section of Wood Preservation conducts studies of various wood preserving processes, of fire-proofing, of glues, and of painting and finishing. It is headed by a Chief of Section, who is a principal chemist, and includes a senior wood technologist, several grades of engineers, a senior chemist, a scientific aid, laboratory aids, and a stenographic and general clerk.

The Section of Industrial Investigations conducts investigations on the utilization of waste for small dimension stock, on logging and milling, on lumber grading, on little used species, and on the passing on of the waste of one industry for use by another. The work of the Section is directed by a Chief of Section, a principal engineer, who is aided by an Assistant Section Chief, and a senior forester. Other employees include wood technologists and engineers of different grades, a senior lumber inspector, and stenographic clerks.

Office of Range Research. The Office of Range Research deals with grazing, rather than timber, problems. It collects, analyzes, and interprets range data, investigates methods of improving range conditions, studies grazing practices, conducts experiments in the elimination of noxious and worthless vegetation, and determines the interrelationships of range use and other forest and range resources.

The general administrative work of the Office is directed by a Chief, who is aided by a plant ecologist. The central office includes a Section of Range Forage Investigations, the work of which is conducted in coöperation with the Bureau of Plant Industry.

The work is carried on in the field at three locations:
1. Great Basin Range Experiment Station, Ogden, Utah
2. Jornada Range Reserve, Las Cruces, New Mexico
3. Santa Rita Range Reserve, Tucson, Arizona

Field Administration. The Field Administration of the Forest Service is organized on a geographical or district basis, the sub-organization, by districts, being functional. The United States, including Alaska and Porto Rico, is divided into nine forest districts as follows:

1. Northern
   Territory: Montana, northeastern Washington, northern Idaho, northwestern South Dakota.
   Headquarters: Missoula.
2. Rocky Mountain
   Territory: Colorado, Wyoming, South Dakota, Nebraska, and western Oklahoma.
   Heaquarters: Denver.
3. Southwestern
   Territory: New Mexico and Arizona.
   Headquarters: Albuquerque.
4. Intermountain
   Territory: Utah, southern Idaho, western Wyoming, Nevada, and northwestern Arizona.
   Headquarters: Ogden.
5. California
   Territory: California and southwestern Nevada.
   Headquarters: San Francisco.
6. North Pacific
   Territory: Washington and Oregon.
   Headquarters: Portland.
7. Eastern
   Territory: Maine, New Hampshire, Vermont, Pennsylvania, Virginia, West Virginia, North Carolina, South Carolina, Georgia, Florida, Alabama, Arkansas, Mississippi, Louisiana, Tennessee, and Porto Rico.
   Headquarters: Washington, D. C.
8. Alaska
   Territory: Alaska.
   Headquarters: Juneau.
9. Lake States
   Territory: Michigan, Minnesota, and Wisconsin.
   Headquarters: Milwaukee.

District Organization. Each of the nine forest districts into which the country is divided is organized along the lines of the headquarters in Washington. The administration of the field work, in other words, is almost completely decentralized, each district having considerable local responsibility.

Each district is under the supervision of a District Forester, who reports directly to headquarters at Washington, but exercises a large degree of independence˙and wide discretionary authority.

A typical district is organized as follows: The District Forester with the necessary clerical assistance forms the Office proper of the District Forester, which is responsible for all work in the district. To carry on the work, the following subdivisions or " offices " have been established:

1. Office of Operation
2. Office of Forest Management
3. Office of Range Management
4. Office of Lands
5. Office of Forest Products
6. Office of Engineering
7. Office of Public Relations
8. Office of Solicitor
9. Office of Finance and Accounts

Office of Operation. The Office of Operation is headed by an Assistant District Forester, who reports directly to the District Forester. It is composed of a general administrative group and a Section of Maintenance. The former includes an assistant district forester, four inspectors (two for fire, one for improvements and trails, and one for state coöperative work), a law enforcement officer, an administrative assistant, and clerks.

The Section of Maintenance is headed by a Chief of Maintenance and Purchasing Agent, who is aided by an Assistant Chief. The remaining personnel includes a bookkeeper, warehouse foremen, assistant warehouse foremen, a telephone operator, a mail clerk, a mimeograph operator, a messenger, and stenographic clerks.

The work of the Office, in general, includes such activities as the following:

Fire control, including suppression and state and private coöperation
Law enforcement
Maintenance
Personnel inspection and training
Improvements
Trails
State coöperation in distribution of forest planting stock
Telephone engineering
Protection of California and Oregon railroad lands [2]

Office of Forest Management. The Office of Forest Management is under direction of an Assistant District Forester, who is aided by a technical assistant. The general staff also includes two logging engineers, a surveyor, and the necessary clerical help. A nursery is also maintained. This is headed by a Chief of Planting, who has an assistant and a clerk.

The Office of Forest Management has supervision over:

Logging
Planting
Coöperative distribution of forest planting stock
Timber surveys
Timber sales
Forest insect investigations [3]
Forest pathology [4]
Timber sales, field supervision

Office of Range Management. An Assistant District Forester heads the Office of Range Management. The staff includes technical assistants and the necessary clerks. The work includes reconnaissance and research. Offices are maintained in each district except District 8, Alaska, and District 9, the Lake States.

Office of Lands. The Office of Lands is headed by an Assistant District Forester. The staff includes an Inspector, a claim examiner, and a clerk. It is concerned with land exchange, land classification and acquisition, mineral examination, and recreation and land entry surveys.

[2] In District 6, only.
[3] Conducted by the Bureau of Entomology.
[4] Conducted by the Office of Forest Pathology, Bureau of Plant Industry.

Office of Forest Products. An Office of Forest Products is found in but two districts (Nos. 1 and 6). The staff includes a Chief of Forest Products, in charge, two or three technical assistants, and a clerk. Field studies of the character indicated by the name are carried on.

Office of Engineering. The Office of Engineering is headed by a District Engineer, who is aided by an Assistant District Engineer, a Project Engineer, numerous surveyors, draftsmen, a blue printer, and clerks. It is concerned with the following subjects:

Maps	Surveys, general
Drafting	Entry surveys
Roads	Water power

Office of Public Relations. The typical Office of Public Relations [5] is headed by an Assistant District Forester, with additional employees, such as an administrative assistant, a clerk and in two districts [6] a librarian. In certain districts (Nos. 2 and 4) this Office is a part of the general " overhead " or administrative machinery. The work is obvious from the title.

Office of Solicitor. At the head of the Office of Solicitor of each of the forest districts is a District Assistant to the Solicitor for the Department of Agriculture, who is aided by a clerk. Districts Nos. 7 and 9 are exceptions. The title of the law officer for District No. 7 is Senior Attorney, Office of the Solicitor, though his duties correspond to those of the law officers of other districts.

In addition, this district office lists an attorney, numerous title attorneys, a special attorney, and an abstractor, engaged in the legal work connected with the acquisition of lands under the Weeks and Clarke-McNary acts. Certain of these title attorneys are also commissioned as special assistants to the United States District Attorneys for the federal judicial districts in which they are located.

[5] There is no such office or unit in Districts 8 and 9.
[6] Districts 5 and 6.

District No. 9 lists only title attorneys for handling its legal work.

In general the District Assistant to the Solicitor transacts, in his district, all legal business arising therein in connection with forests. In addition he conducts any other Department of Agriculture legal work which may be necessary.

Office of Finance and Accounts. The District Office of Finance and Accounts is under the direction of a District Fiscal Agent, who is aided by a deputy.' The additional personnel includes such employees as an auditor, bookkeeper, assistant bookkeeper, check writer, stenographer, property audit clerk, and clerk. The district accounting and auditing functions are completely decentralized, each district organization being independent save for a yearly audit from Washington.

National Forests. Each of the nine forest districts is subdivided into a number of national forests. The directing officer of each forest is known as the Forest Supervisor, his duty being to direct operations in the forest under his supervision, including management, operation, grazing, etc. To assist him in his work, he has, in addition to the necessary clerical help, an assistant supervisor, when the volume of work justifies it, and one or more technical assistants, who are variously termed junior foresters, and forest examiners. These assistants are employed in various kinds of technical and administrative work.

The national forests, by districts, with headquarters, are as follows:

District	Name	Location	Headquarters
1	Absaroka	Montana	Livingston, Montana
	Beartooth	"	Billings, Montana
	Beaverhead	"	Dillon, Montana
	Bitterroot	"	Hamilton, Montana
	Blackfeet	"	Kalispell, Montana
	Cabinet	"	Thompson Falls, Montana
	Clearwater	Idaho	Orofino, Idaho
	Cœur d'Alene	"	Cœur d'Alene, Idaho
	Custer	Montana-South Dakota	Miles City, Montana

' Except in District 8.

Dis-trict	Name	Location	Headquarters
	Deerlodge	Montana	Butte, Montana
	Flathead	"	Kalispell, Montana
	Gallatin	"	Bozeman, Montana
	Helena	"	Helena, Montana
	Jefferson	"	Great Falls, Montana
	Kaniksu	Idaho-Washington	Newport, Washington
	Kootenai	Montana	Libby, Montana
	Lewis and Clark	"	Choteau, Montana
	Lolo	"	Missoula, Montana
	Madison	"	Sheridan, Montana
	Missoula	"	Missoula, Montana
	Nezperce	Idaho	Grangeville, Idaho
	Pend Oreille	"	Sandpoint, Idaho
	St. Joe	"	St. Maries, Idaho
	Selway	"	Kooskia, Idaho
2	Arapaho	Colorado	Hot Sulphur Springs, Colorado
	Bighorn	Wyoming	Sheridan, Wyoming
	Black Hills	South Dakota-Wyoming	Deadwood, South Dakota
	Cochetopa	Colorado	Salida, Colorado
	Colorado	"	Fort Collins, Colorado
	Grand Mesa	"	Grand Junction, Colorado
	Gunnison	"	Gunnison, Colorado
	Harney	South Dakota	Custer, South Dakota
	Holy Cross	Colorado	Glenwood Springs, Colorado
	Medicine Bow	Wyoming	Laramie, Wyoming
	Montezuma	Colorado	Mancos, Colorado
	Nebraska	Nebraska	Halsey, Nebraska
	Pike	Colorado	Colorado Springs, Colorado
	Rio Grande	"	Monte Vista, Colorado
	Routt	"	Steamboat Springs, Colorado
	San Isabel	"	Pueblo, Colorado
	San Juan	"	Durango, Colorado
	Shoshone	Wyoming	Cody, Wyoming
	Uncompahgre	Colorado	Delta, Colorado
	Washakie	Wyoming	Lander, Wyoming
	White River	Colorado	Glenwood Springs, Colorado
	Wichita	Oklahoma	Cache, Oklahoma
3	Apache	Arizona-New Mexico	Springerville, Arizona
	Carson	New Mexico	Taos, New Mexico
	Coconino	Arizona	Flagstaff, Arizona
	Coronado	New Mexico-Arizona	Tucson, Arizona
	Crook	Arizona	Safford, Arizona
	Datil	New Mexico	Magdalena, New Mexico
	Gila	"	Silver City, New Mexico
	Lincoln	"	Alamogordo, New Mexico
	Manzano	"	Albuquerque, New Mexico
	Prescott	Arizona	Prescott, Arizona
	Santa Fe	New Mexico	Santa Fe, New Mexico
	Sitgreaves	Arizona	Holbrook, Arizona
	Tonto	"	Phœnix, Arizona
	Tusayan	"	Williams, Arizona

9

District	*Name*	*Location*	*Headquarters*
4	Ashley	Utah-Wyoming	Vernal, Utah
	Boise	Idaho	Boise, Idaho
	Cache	Idaho-Utah	Logan, Utah
	Caribou	Idaho-Wyoming	Montpelier, Idaho
	Challis	Idaho	Challis, Idaho
	Dixie	Nevada-Utah	Cedar City, Utah
	Fishlake	Utah	Richfield, Utah
	Humboldt	Nevada	Elko, Nevada
	Idaho	Idaho	McCall, Idaho
	Kaibab	Arizona	Kanab, Utah
	La Sal	Colorado-Utah	Moab, Utah
	Lemhi	Idaho	Mackay, Idaho
	Manti	Utah	Ephraim, Utah
	Minidoka	Idaho-Utah	Burley, Idaho
	Nevada	Nevada	Ely, Nevada
	Payette	Idaho	Emmett, Idaho
	Powell	Utah	Panguitch, Utah
	Salmon	Idaho	Salmon, Idaho
	Sawtooth	"	Hailey, Idaho
	Targhee	Idaho-Wyoming	St. Anthony, Idaho
	Teton	Wyoming	Jackson, Wyoming
	Toiyabe	Nevada	Austin, Nevada
	Uinta	Utah	Provo, Utah
	Wasatch	"	Salt Lake City, Utah
	Weiser	Idaho	Weiser, Idaho
	Wyoming	Wyoming	Kemmerer, Wyoming
5	Angeles	California	Los Angeles, California
	California	"	Willows, California
	Cleveland	"	San Diego, California
	Eldorado	California-Nevada	Placerville, California
	Inyo	"	Bishop, California
	Klamath	California-Oregon	Yreka, California
	Lassen	California	Susanville, California
	Modoc	"	Alturas, California
	Mono	California-Nevada	Minden, Nevada
	Plumas	California	Quincy, California
	San Bernardino	"	San Bernardino, California
	Santa Barbara	"	Santa Barbara, California
	Sequoia	"	Porterville, California
	Shasta	"	Mount Shasta, California
	Sierra	"	North Fork, California
	Stanislaus	"	Sonora, California
	Tahoe	California-Nevada	Nevada City, California
	Trinity	California	Weaverville, California
6	Cascade	Oregon	Eugene, Oregon
	Chelan	Washington	Okanogan, Washington
	Columbia	"	Vancouver, Washington
	Colville	"	Republic, Washington
	Crater	California-Oregon	Medford, Oregon
	Deschutes	Oregon	Bend, Oregon
	Fremont	"	Lakeview, Oregon
	Malheur	"	John Day, Oregon

Dis-trict	Name	Location	Headquarters
	Mount Baker	Washington	Bellingham, Washington
	Mount Hood	Oregon	Portland, Oregon
	Ochoco	"	Prineville, Oregon
	Olympic	Washington	Olympia, Washington
	Rainier	"	Tacoma, Washington
	Santiam	Oregon	Albany, Oregon
	Siskiyou	California-Oregon	Grants Pass, Oregon
	Siuslaw	Oregon	Eugene, Oregon
	Snoqualmie	Washington	Seattle, Washington
	Umatilla	Oregon-Washington	Pendleton, Oregon
	Umpqua	Oregon	Roseburg, Oregon.
	Wallowa	"	Wallowa, Oregon
	Wenatchee	Washington	Wenatchee, Washington
	Whitman	Oregon	Baker, Oregon
7	Alabama	Alabama	Athens, Tennessee
	Allegheny	Pennsylvania	Warren, Pennsylvania
	Cherokee	Georgia-North Carolina-Tennessee	Athens, Tennessee
	Choctawhatchee	Florida	Pensacola, Florida
	Luquillo	Porto Rico	Rio Piedras, Porto Rico
	Monongahela	Virginia-West Virginia	Elkins, West Virginia
	Nantahala	South Carolina-North Carolina-Geogia	Franklin, North Carolina
	Natural Bridge	Virginia	Lynchburg, Virginia
	Ocala	Florida	Lake City, Florida
	Ouachita	Arkansas	Hot Springs National Park, Arkansas
	Ozark	"	Russellville, Arkansas
	Pisgah	North Carolina-Tennessee	Asheville, North Carolina
	Shenandoah	Virginia-West Virginia	Harrisonburg, Virginia
	Unaka	North Carolina-Tennessee-Virginia	Bristol, Tennessee
	White Mountain	Maine-New-Hampshire	Laconia, New Hampshire
8	Chugach	Alaska	Cordova, Alaska
	Tongass	"	Ketchikan, Alaska
9	Chippewa	Minnesota	Cass Lake, Minnesota
	Huron	Michigan	East Tawas, Michigan
	Marquette	Michigan	Munising, Michigan
	Superior	Minnesota	Ely, Minnesota

The total area in acres was 159,750,520 at the end of the fiscal year 1929.

Monuments. Certain national monuments are situated within the national forests and are administered by the Department of Agriculture, through the Forest Service. These are:

Name	Forest	Location	Areas
Bandelier	Santa Fe	New Mexico	22,075
Chiricahua	Coronado	Arizona	4,480
Devil Postpile	Sierra	California	800
Gila Cliff Dwellings	Gila	New Mexico	160
Holy Cross	Holy Cross	Colorado	1,392
Jewel Cave	Harney	South Dakota	1,280
Lava Beds	Modoc	California	45,967
Lehman Caves	Nevada	Nevada	593
Mount Olympus	Olympic	Washington	298,730
Old Kasaan	Tongass	Alaska	38
Oregon Caves	Siskiyou	Oregon	480
Timpanogos Cave	Wasatch	Utah	250
Tonto	Tonto	Arizona	640
Walnut Canyon	Coconino	Arizona	960
Wheeler	Cochetopa, Rio Grande	Colorado	300
		Total	378,145

Game Refuges. The following national game refuges, situated wholly or in part within national forests, have been designated for the protection of game:

Name	Forest	Location	Areas
Cherokee National Game Refuge No. 1	Cherokee	Tennessee	30,000
Cherokee National Game Refuge No. 2	Cherokee	Georgia	14,000
Custer State Park Game Sanctuary	Harney	South Dakota	44,840
Grand Canyon	Tusayan (Kaibab)	Arizona	792,163
Ozark National Game Refuge No. 1	Ozark	Arkansas	8,420
Ozark National Game Refuge No. 2	Ozark	Arkansas	5,300
Ozark National Game Refuge No. 3	Ozark	Arkansas	3,620
Ozark National Game Refuge No. 4	Ozark	Arkansas	4,160
Pisgah	Pisgah	North Carolina	98,381
Sequoia	Sequoia	California	16,300
Sheep Mountain	Medicine Bow	Wyoming	28,318
Wichita	Wichita	Oklahoma	60,800
	Black Hills (Meade District)	South Dakota	5,548
	Manzano (Zuni District)	New Mexico	45,423
	Medicine Bow (Pole Mountain District)	Wyoming	56,132
	Michigan (Brady District)	Michigan	2,680

Range Reserves. In addition there have been established, by executive order for use of the Forest Service in conducting studies of grazing and range management, two range reserves; Jornada (193,686 acres) in New Mexico, and Santa Rita (52,399 acres) in Arizona.

Ranger Districts. Each forest, in turn, is divided into several ranger districts, each managed by a Forest Ranger, who is agent of the Forest Supervisor in carrying on the primary field, or technical work of the Forest Service. If the volume of work justifies it, the Ranger has an assistant, on full time, throughout the year. In addition there are temporary employees, known as fire guards, under direction of the Ranger, during the season of greatest fire hazard.[8]

Besides this emergency rank and file, there are emergency directive units in regions of high hazard in the Pacific Northwest. These units, which are a part of the district organization and devised to be moved to any point within the district as emergencies require, consist of a forest-fire chief, who is a regular ranger with long experience in handling large fires, and a squadron composed of two or three fire foremen, a camp superintendent, and a cook—all local residents with fire-fighting experience. These units are stationed at strategic points throughout the district, supplied with motor transport, and are in constant telephone touch with Forest Supervisors' headquarters.

The ranger districts may number from two to ten, depending upon the forest and the ease of communication. In addition to these districts, numerous forests have other administrative subdivisions which are functional, rather than geographical. These include such titles as the following: Timber Sale, Timber Surveys, Eagle Creek Forest Camp, and those named after national monuments or reservations.

Miscellaneous. In addition to the complex and wide spread organization just described there are three boards or commissions

---

[8] In the Service, as a whole, some three thousand of these temporary employees are taken on during the summer months.

which, while they are in no sense a part of the Forest Service, are connected with its work.

National Forest Reservation Commission. The National Forest Reservation Commission is composed of the Secretaries of War, the Interior, and Agriculture, two Senators, two Representatives, and a Secretary. This Commission is charged with the responsibility of approving for acquisition, under the Weeks and Clarke-McNary acts, lands suitable for incorporation into the national forests.

Forest Protection Board. The Forest Protection Board is composed of representatives of the National Park Service, the General Land Office, the Office of Indian Affairs, the Weather Bureau, the Bureau of Entomology, the Bureau of Plant Industry, the Bureau of Biological Survey, and the Forest Service.

The purpose of this board is to formulate and recommend to this office [Chief Coördinator] general policies and plans for the protection of the forests of the country, especially in the prevention and suppression of forest fires, embracing measures for the coöperation of Federal, State and private agencies.

The Board functions, in handling the local problems of interbureau coöperation in forest protection, through committees chosen from the field personnel of the eight services involved, there being six committees in all, one for each of the Forest Service districts in the West.

National Capital Park and Planning Commission. The National Capital Park and Planning Commission was established " to develop a comprehensive, consistent and coördinated plan for the National Capital and its environs in the States of Maryland and Virginia, to preserve the flow of water in Rock Creek, to prevent pollution of Rock Creek and the Potomac and Anacostia Rivers, to preserve forests and natural scenery in and about Washington . . . etc." While it bears no direct relation to the Forest Service, the Forester is a member of the Commission. Certain problems arise, from time to time, which fall within the technical purview of the Forest Service.

APPENDIX 1

OUTLINE OF ORGANIZATION

Explanatory Note

The purpose of the Outlines of Organization in this series of Monographs is to show in detail the organization and personnel of the several services of the national government to which they relate. They have been prepared in accordance with the plan followed by the President's Commission on Economy and Efficiency in its outlines of the organization of the United States Government.[1] They differ from those outlines, however, in that whereas the Commission's report showed only organization units, the presentation herein has been carried far enough to show the personnel embraced in each organization unit.

These outlines are of value not merely as an effective means of making known the organization of the several services. If kept revised to date, they constitute exceedingly important tools of administration. They permit the directing personnel to see at a glance the organization and personnel at their disposal. They establish definitely the line of administrative authority and enable each employee to know his place in the system. They furnish the essential basis of plans for determining costs by organization division and sub-division. They afford the data for a consideration of the problem of classifying and standardizing personnel and compensation. Collectively they make it possible to determine the number and location of organization units of any particular kind, such as, laboratories, libraries, blue-print rooms, or other plants, to what services attached and where located, or to determine what services are maintaining stations at any city or point in the United States. The Institute hopes that

[1] 62 Cong., H. doc. 458, 1912, 2 vols.

123

upon the completion of the present series it will be able to prepare a complete classified statement of the technical and other facilities at the disposal of the government. The present monographs will then furnish the details regarding the organization, equipment, and work of the institutions so listed and classified.

OUTLINE OF ORGANIZATION

FOREST SERVICE

DEPARTMENT OF AGRICULTURE

NOVEMBER, 1929

Units of Organization; Classes of Employees	Number	Annual Salary Rate
1. General Administration		
1. Office of the Forester		
The Forester (Chief of Bureau)	1	$8,000
Associate Forester	1	7,000
Clerk (Secretary)	1	2,600
Clerk (Stenographer)	1	2,040
2. Branch of Finance and Accounts		
Chief of Finance and Accounts	1	5,200
Assistant Chief	1	3,700
Accountant	2	3,000
Appointment Clerk	1	2,500
Transportation and Rate Clerk	1	2,300
Auditor	1	2,300
Clerk (Stenographer)	1	1,980
Clerk (Bookkeeper)	1	1,920
Clerk	1	1,800
Clerk (Stenographer)	1	1,620
3. Branch of Operation		
1. Office of the Chief of Branch		
Assistant Forester	1	6,000
Assistant Chief of Branch	1	4,800
Inspector	1	5,000
Administrative Assistant	1	3,700
Clerk (Branch Files)	1	1,980
Clerk (Ediphone Operator)	1	1,800
2. Office of Maintenance		
1. Office of the Chief of Maintenance		
Chief of Maintenance	1	2,400
Clerk (Purchasing Agent)	1	1,980
Clerk (Mail)	1	1,920
Clerk (Stenographer)	1	1,620
Clerk (Telephone Operator)	1	1,380
Electrician	1	1,800
Carpenter	1	1,740

Messenger (Head)	1	1,380
Messenger	2	1,320
	1	1,260
	3	1,200
	1	1,140
	2	720
	1	600
Laborer	1	1,260
	1	1,080
Charwoman (Head)	1	1,080
Charwoman (Part time)	9	413

2. Section of Stenography and Typing

Clerk (Chief of Section)	1	2,200
Clerk (Stenographer and Typist)	1	1,800
	2	1,560
Clerk (Typist)	1	1,740
	1	1,620
	2	1,440
Mimeograph Operator	1	1,440

3. Section of Supplies

Clerk	1	1,620
Laborer	1	1,560
	1	1,380
	1	1,260

3. Supply Depot, Ogden, Utah

Supply Officer	1	3,800
Printer	1	2,100
Clerk	1	2,100
	1	1,860
Clerk (Stenographer)	1	1,680
Instrument-Repairman	1	2,000
Mechanic (Print shop)	1	1,740
Packer	1	1,560
	1	1,440
	2	1,380

4. Branch of Forest Management

Assistant Forester	1	6,000
Assistant Chief of Branch	1	4,800
Forest Inspector	1	4,800
Clerk (Stenographer)	1	1,980
	1	1,920
Clerk (Branch files)	1	1,980

5. Branch of Range Management

Assistant Forester	1	6,000
Inspector	1	4,600
Clerk (Stenographer)	1	1,800

6. Branch of Lands

Assistant Forester	1	6,000
Senior Attorney	1	6,000
Assistant Chief of Branch	1	4,800

Administrative Assistant	1	2,900
Clerk (Land titles)	1	2,300
	1	1,740
Clerk (Stenographer)	1	2,040
Clerk (Branch files)	1	1,980
Collaborator	1	300

7. Branch of Engineering
  1. Office of the Chief of Branch

Chief Engineer	1	6,000
Assistant Chief Engineer	1	5,000
Assistant Engineer	1	3,500
Clerk (Statistical)	1	2,000
Clerk (Stenographer)	1	1,980
	1	1,800
	1	1,740
Clerk (Branch files)	1	1,920

  2. Office of Maps and Surveys
    1. Office of the Assistant Engineer

Assistant Engineer	1	3,200

    2. Section of Drafting

Chief Draftsman	1	2,900
Draftsman	2	2,700
	2	2,600
	2	2,500
	3	2,400
	1	2,300
	1	1,980
	1	1,920
	1	1,860
	3	1,620
Artist	1	2,200
Clerk (Files)	1	1,980

    3. Section of Atlas

Statistician	1	2,700
Draftsman	1	1,860
Clerk	1	1,620

    4. Section of Lithographing

Executive Assistant	1	2,900

    5. Section of Photography

Chief Photographer	1	2,700
Photographer	3	2,200
	1	2,100
	2	1,800
Clerk	1	1,980

8. Branch of Public Relations
  1. Office of the Chief of Branch

Assistant Forester	1	6,000

Assistant Forester (Special assignments)	1	5,000
Logging Engineer (Coöperation with timberland owners)	1	4,800
Clerk (Branch files)	1	2,400
	1	1,980
Clerk (Secretary)	1	2,040

2. Division of State Coöperation
  1. Office of the Chief of Division

Forest Inspector (Chief of Division)	1	5,000
Forest Inspector	1	4,800
Law Compiler	1	1,800
Clerk (Stenographer)	1	1,860
	2	1,620

  2. Fire Protection and Distribution of Planting Stock
    1. Northeastern States, Amherst, Massachusetts

District Inspector	1	4,800
Clerk (half time)	1	1,740

    2. Middle Atlantic States, Washington, D. C.

District Inspector	1	4,600

    3. Southeastern States, Asheville, North Carolina

District Inspector	1	4,600
Clerk (half time)	1	1,620

    4. Gulf States, New Orleans, Louisiana

District Inspector	1	4,600
Clerk (half time)	1	1,620

    5. Central States, Louisville, Kentucky

District Inspector	1	4,600
Clerk	1	1,800

  3. Farm Forestry

Extension Forester	1	4,800
Clerk (Stenographer)	1	1,680

3. Division of Information
  1. Office of the Chief of Division

Chief of Division	1	5,000
Editor (Press relations)	1	3,200
Artist	1	2,300
Clerk	1	2,040
Clerk (Dictaphone Operator)	1	1,860

  2. Office of Educational Coöperation

Editor	1	3,900
Clerk (Editorial)	1	2,500
	1	2,400
Clerk (Correspondence)	1	1,920
Clerk	1	1,740

3. Office of Publication
   1. Office of the Chief

Chief of Publication	1	3,300
Clerk (Stenographer)	1	1,680

   2. Section of Printing

Clerk (Chief of Section)	1	2,600
Clerk	1	1,980
	1	1,800
	1	1,440

4. Office of Visual Education

Chief of Office	1	3,600
Clerk	1	1,920
	1	1,680
Clerk (Stenographer)	1	1,560
Carpenter	1	1,500

9. Branch of Research
   1. Office of the Chief of Branch

Assistant Forester	1	6,000
Assistant Chief of Branch	1	5,200
Editor	1	3,700
Administrative Assistant	1	3,700
Clerk (Branch files)	1	2,400
Clerk (Secretary)	1	2,040
Clerk (Assistant in Branch files)	1	1,920
Clerk (Compilation File)	1	1,860

  2. Office of Silvics
   1. Office of the Chief

Chief of Office	1	5,000
Technical Assistant	1	3,500
	1	3,400
Clerk (Stenographer)	1	1,980
	1	1,920

   2. Section of Forest Measurements

Chief of Section	1	3,000
Silviculturist	1	4,800
Technical Assistant	1	2,600
	1	2,100
Clerk (Statistical)	1	2,200
	2	2,000
	1	1,860
	1	1,800
	1	1,680
	1	1,620
Clerk (Tabulating Machine)	1	1,740
	1	1,620

   3. Section of Library

Librarian	1	2,600

4. Field Division
  1. Allegheny Forest Experiment Station,
     Philadelphia, Pennsylvania

Director	1	4,000
Investigative Assistant	1	3,200
	2	2,600
	1	2,200
	2	2,000
Clerk	1	2,000
	2	1,620

  2. Appalachian Forest Experiment Sta-
     tion, Asheville, North Carolina

Director	1	4,800
Investigative Assistant	1	4,600
	1	3,800
	1	2,700
	1	2,300
	1	2,100
	1	2,000
Clerk	1	2,000
	1	1,800

  3. California Forest Experiment Station
     Berkeley, California

Director	1	5,000
Investigative Assistant	1	4,800
	2	4,600
	1	4,000
	2	3,400
	1	3,200
	1	2,600
	1	1,800
Clerk	1	2,100
	2	1,800
	2	1,440

  4. Central States Forest Experiment
     Station, Columbus, Ohio

Director	1	4,800
Investigative Assistant	1	4,000
	2	2,600
	1	2,400
	1	2,000
Clerk	1	2,000
	1	1,440

  5. Lake States Forest Experiment
     Station, St. Paul, Minnesota

Director	1	6,000

Investigative Assistant	1	5,000
	1	4,800
	1	3,800
	1	3,600
	1	2,400
	6	2,000
Clerk	1	2,300
	1	1,800
	1	1,440

6. Northeastern Forest Experiment
   Station, Amherst, Massachusetts

Director	1	6,000
Investigative Assistant	1	3,800
	1	2,800
	1	2,500
Clerk	1	2,200
	1	1,800
	1	1,620
Clerk (half time)	1	1,740

7. Northern Rocky Mountain Forest
   Experiment Station,
   Missoula, Montana

Director	1	4,600
Investigative Assistant	1	3,800
	1	2,800
	1	2,600
	1	2,200
Clerk	1	1,800

8. Pacific Northwest Forest Experiment
   Station, Portland, Oregon

Director	1	5,200
Investigative Assistant	1	3,200
	2	2,800
	1	2,600
	2	2,000
	1	2,040
Clerk	1	2,300
	1	1,560

9. Rocky Mountain Forest Experiment
   Station, Colorado Springs, Colorado

Investigative Assistant (in charge)	1	2,800

10. Southern Forest Experiment Station,
    New Orleans, Louisiana
    1. Office of Director

Director	1	4,600

Investigative Assistant	1	5,000
	1	3,300
	1	2,700
	1	2,100
	4	2,000
Clerk	1	2,600
	1	2,200
	1	1,980
	2	1,620
Clerk (half-time)	1	1,620
	1	1,440

2. Sub-station, Starke, Florida

Investigative Assistant (in charge)	1	3,500
Investigative Assistant	1	2,100
Woods Superintendent	1	1,620
Clerk	1	1,440

11. Southwestern Forest Experiment Station, Flagstaff, Arizona

Director	1	5,000
Investigative Assistant	1	3,300
	1	1,800
Clerk	1	2,000
Laborer	1	1,800

3. Office of Forest Economics
   1. Office of Senior Economist

Senior Economist	1	5,200
Clerk (Stenographer)	1	1,680

   2. Economic Studies

Forest Economist	1	4,800
	1	3,500
	1	3,400
Statistician in Forest Products	1	2,600
Technical Assistant	1	2,300
Clerk (Statistical)	1	2,200
Clerk (Stenographer)	1	1,680
Clerk	1	1,620

   3. Forest Taxation Inquiry, New Haven, Connecticut

Economist (in charge)	1	8,000
Investigative Assistant	2	5,600
	1	4,600
	2	4,000
	1	2,300
Mathematician	1	2,600

Clerk	1	2,100
	1	1,800
	2	1,620
	1	1,500

4. Office of Forest Products

Engineer in Forest Products	1	4,800
Associate Scientific Assistant	2	3,400
Clerk (Stenographer)	1	1,920
	1	1,740

5. Forest Products Laboratory, Madison, Wisconsin

  1. Office of the Director

Director	1	6,000
Principal Chemist	1	6,000
Senior Clerk (Secretary)	1	2,200

  2. Office of Dry Kiln Expert

Physicist and Expert in Kiln Drying	1	5,000

  3. Section of Publication of Results

Senior Forester (in charge)	1	4,800
Senior Technical Reviewer	1	4,600
Technical Reviewer	1	3,800
Forester	1	3,800
Assistant Technical Writer	1	2,600
Multigraph Foreman	1	2,040
Clerk (Typist and General)	1	1,800
Assistant Clerk (Stenographer)	1	1,620

  4. Section of Silvicultural Relations

Specialist in Wood Structure (in charge)	1	5,200
Microscopist in Forest Products	1	4,800
Silviculturist	1	3,800
Assistant Physiological Plant Anatomist	1	2,600
Assistant Wood Technologist	1	2,600
Junior Xylotomist	1	2,200
Assistant Laboratory Aid (Translator)	1	1,620

  5. Section of Derived Products

Chief of Section (Chemist)	1	5,400
Chemist	2	5,000
	1	4,200
Chemist (Colloidal)	1	3,800
Associate Chemist	1	3,200
Assistant Chemist	1	2,700
Junior Chemist	2	2,300
	1	2,200
	3	2,000
Clerk (Stenographer and General)	1	1,800
Under Laboratory Aid	1	1,260

6. Section of Pulp and Paper

Chief of Section (Chemical Engineer)	1	5,800
Chemist (Assistant Section Chief)	1	5,000
Senior Chemist in Forest Products	1	4,600
Chemist in Forest Products	1	3,800
Associate Chemist in Forest Products	1	3,200
Assistant Chemist	1	2,600
Junior Chemist	1	2,000
	3	2,000
Associate Engineer in Forest Products	1	3,400
Assistant Engineer in Forest Products	1	2,800
Assistant Engineer (Chemical)	1	2,800
Assistant Engineer	1	2,600
Papermaker	1	1,980
Laboratory Aid	2	1,800
Assistant Laboratory Aid	2	1,680
Clerk (Stenographer and General)	1	1,800
Assistant Clerk (Stenographer)	1	1,620
Mechanics	1	1,680

7. Section of Timber Mechanics

Chief of Section (Engineer)	1	5,800
Senior Engineer in Forest Products	2	5,000
	1	4,800
	1	4,600
Engineer in Forest Products	1	3,800
Associate Engineer in Forest Products	1	3,500
	2	3,200
Assistant Engineer in Forest Products	1	2,700
Senior Laboratory Aid	1	2,300
Laboratory Aid	1	1,860
Assistant Laboratory Aid	1	1,680
	2	1,620
Senior Clerk (Statistical)	1	2,100
Clerk (Stenographer)	1	1,980
Assistant Clerk (Stenographer)	2	1,620
Carpenter	1	1,800

8. Section of Timber Physics

Chief of Section (Engineer)	1	5,600
Senior Engineer (Assistant Section Chief)	1	4,600
Senior Engineer in Forest Products	2	4,600
Engineer in Forest Products	1	4,200
Associate Engineer	1	3,300
Assistant Engineer	1	2,800
Associate Wood Technologist	1	3,300
Assistant Wood Technologist	1	2,700
Laboratory Aid	2	1,920

Junior Laboratory Aid	1	1,500
Clerk (Stenographer and General)	1	1,860

9. Section of Wood Preservation

Chief of Section (Chemist)	1	5,600
Senior Wood Technologist	1	4,800
Senior Chemist in Forest Products	1	4,600
Junior Chemist in Forest Products	1	2,000
Engineer in Forest Products	1	4,600
Associate Engineer in Forest Products	1	3,300
Assistant Engineer in Forest Products	2	2,800
Assistant Engineer (Chemicals)	1	2,600
Junior Engineer	1	2,300
Scientific Aid	1	1,920
Clerk (Stenographer and General)	1	1,860
Assistant Laboratory Aid	2	1,620

10. Section of Industrial Investigations

Chief of Section (Engineer)	1	5,200
Forester (Assistant Section Chief)	1	4,800
Forester	1	4,000
Engineer in Forest Products	1	3,800
Associate Engineer in Forest Products	1	3,200
Wood Technologist	2	4,000
Associate Wood Technologist	1	3,300
	1	3,200
Senior Lumber Inspector	1	2,200
Clerk (Stenographer and General)	1	1,800
Assistant Clerk (Stenographer)	1	1,620

11. Section of Pathology (in coöperation with Bureau of Plant Industry, Department of Agriculture)

Clerk (Stenographer and General)	1	1,800

12. Section of Laboratory Operation

Chief of Section (Engineer)	1	4,800
Associate Engineer in Forest Products	1	3,300
Associate Chemist (Photography)	1	3,200
Senior Administrative Assistant	1	3,300
Librarian	1	2,600
Senior Library Assistant	1	2,200
Building and Property Custodian	1	2,400
Principal Clerk (Computing)	1	2,300
Senior Clerk (Purchasing)	1	2,000
Clerk (Stenographer and General)	2	1,800
Clerk (Library, File and Mail)	1	1,800
Assistant Clerk (Computing)	1	1,800
	1	1,740
	1	1,680
	2	1,620

Assistant Clerk-Stenographer	1	1,620
Assistant Clerk (Purchasing)	1	1,620
Assistant Clerk (Price and Record)	1	1,620
Junior Clerk-Typist	2	1,440
Junior Clerk (Stenographer)	2	1,440
Junior Clerk	1	1,320
Under Clerk	1	1,320
Under Clerk (Storekeeper)	1	1,200
Carpenter Foreman	1	2,400
Carpenter	1	2,100
	4	1,980
	1	1,740
Draftsman	1	2,300
Photographer	1	2,040
Machine Shop Foreman	1	2,200
Senior Mechanic	1	1,980
Senior Mechanic (Electrician)	1	1,920
Junior Mechanic (Carpenter)	1	1,560
Typewriter Repairman	1	1,800
Skilled Laborer	1	1,500
Assistant Laboratory Aid	1	1,620
Under Operator (Card Punch Machine)	1	1,380
	3	1,260
Office Appliance Operator	1	1,260
Telephone Operator	1	1,260
Janitor	1	1,440
	3	1,380
	1	1,260
	1	1,200
Watchman	1	1,200
	1 (per d.)	3.83
Messenger	1	1,140
	1	1,080
	1	840
Charwoman	3 (per h.)	.45
13. Section of Finance and Accounts		
Fiscal Agent (Section Chief)	1	3,800
Deputy Fiscal Agent	1	2,800
Time and Cost-Property Clerk	1	2,100
Clerk (Stenographer and General)	1	1,800
6. Office of Range Research		
1. Office of the Inspector		
Inspector (in charge)	1	4,800
Clerk (Stenographer)	1	1,920
2. Range Forage Investigations		
Plant Ecologist	1	3,800
Assistant Botanist (part time)	1	2,800

Junior Plant Ecologist	1	2,200
Clerk	1	1,800
Forage Plant Mounter	1	1,620

3. Great Basin Range Experiment Station, Ogden, Utah

Director	1	5,000
Investigative Assistant	1	4,600
	1	3,400
	1	3,200
	1	2,600
Clerk	1	2,200
	1	2,000
	1	1,800

4. Jornada Range Reserve, Las Cruces, New Mexico

Investigative Assistant in charge	1	2,700
Investigative Assistant	1	2,200

5 Santa Rita Range Reserve, Tucson, Arizona

Director	1	3,400
Investigative Assistant	1	2,200

2. Field Administration

1. National Forest District No. 1—Northern District [1]

1. District Headquarters, Missoula, Montana

District Forester	1	6,000
Clerk (Stenographer)	1	1,980

2. Office of Finance and Accounts

District Fiscal Agent	1	4,600
Deputy Fiscal Agent	1	3,200
Clerk (Bookkeeper)	1	2,500
Clerk (Auditor)	1	2,400
Clerk (Assistant Bookkeeper)	1	2,100
Clerk	1	1,980
Clerk (Property Audit)	1	1,860
Clerk (Stenographer)	1	1,680
Clerk (Check writer)	1	1,680

[1] The administrative units and personnel in this District headquarters office are listed in full for purposes of illustration. The organizations of the eight other District headquarters offices do not differ essentially from that of the Northern District, the number of employees of the different grades varying only as this may be necessitated by differences in the predominating National Forest interests in the several regions. The organization for direct forestry work in the forests varies as seasons and conditions demand; hence no attempt has been made to present a typical set-up in this outline. For description of typical forest organization see text, p. 116.

3. Office of Operation
   1. Office of the Assistant District Forester

Assistant District Forester	1	4,800
Inspector (Fire Control)	1	4,800
	1	4,600
Inspector (Improvements and Trails)	1	3,800
Inspector (State coöperation)	1	3,600
Administrative Assistant	1	3,500
Law Enforcement Officer	1	3,400
Clerk (Stenographer and Files)	1	1,860

   2. Section of Maintenance

Chief of Maintenance and Purchasing Agent	1	3,700
Assistant Chief	1	2,600
Warehouse Foreman (Spokane warehouse)	1	2,300
Warehouse Foreman (Missoula warehouse)	1	1,980
Assistant Warehouse Foreman	1	1,860
	2	1,680
Clerk (in charge Stenographic Section)	1	1,860
Clerk (Bookkeeper)	1	1,800
Clerk (Stenographer)	1	1,980
	1	1,800
	1	1,680
	1	1,620
	2	1,500
	1	1,440
Mimeograph Operator	1	1,380
Telephone Operator	1	1,380
Mail Clerk	1	1,320
Messenger	1	600

4. Office of Forest Management
   1. Office of the Assistant District Forester

Assistant District Forester	1	4,800
Technical Assistant	1	4,000
Logging Engineer	1	4,600
	1	3,600
Surveyor	1	2,700
Clerk (Stenographer and Files)	1	1,860
Clerk (Librarian)	1	1,620

   2. Savenac Nursery, Haugen, Montana

Chief of Planting	1	3,600
Assistant in Planting	1	2,900
Clerk (General)	1	1,860

5. Office of Range Management

Assistant District Forester	1	4,800
Inspector	1	3,600
Technical Assistant	1	1,800
Clerk (General)	1	1,800

6. Office of Lands

Assistant District Forester	1	4,600
Inspector	1	3,800
Claims Examiner	1	2,600
Clerk (General)	1	1,860

7. Office of Forest Products

Chief of Forest Products	1	4,600
Technical Assistant	1	3,300
	1	2,800
Clerk (General)	1	1,860

8. Office of Engineering

District Engineer	1	4,800
Assistant District Engineer	1	3,600
Project Engineer	1	3,600
Surveyor	1	3,100
	2	2,800
	1	2,700
	1	2,600
	1	2,500
Chief Draftsman	1	3,000
Draftsman	1	2,500
	2	2,400
	1	2,100
	1	1,620
Clerk (General)	1	2,200
Clerk (Stenographer)	1	1,800
	1	1,740

9. Office of Public Relations

Assistant District Forester	1	4,600
Administrative Assistant	1	3,000
Clerk (General)	1	1,800

10. Office of Solicitor

District Assistant to Solicitor	1	4,600
Clerk	1	1,860

APPENDIX 2

CLASSIFICATION OF ACTIVITIES

Explanatory Note

The Classifications of Activities in this series have for their purpose to list and classify in all practicable detail the specific activities engaged in by the several services of the national government. Such statements are of value from a number of standpoints. They furnish, in the first place, the most effective showing that can be made in brief compass of the character of the work performed. Secondly, they lay the basis for a system of accounting and reporting that will permit the showing of total expenditures classified according to activities. Finally, taken collectively, they make possible the preparation of a general or consolidated statement of the activities of the government as a whole. Such a statement will reveal in detail, not only what the government is doing, but the services in which the work is being performed. For example, one class of activities that would probably appear in such a classification is that of " scientific research." A subhead under this class would be " chemical research." Under this head would appear the specific lines of investigation under way and the services in which they were being prosecuted. It is hardly necessary to point out the value of such information in planning for future work and in considering the problem of the better distribution and coördination of the work of the government. The Institute contemplates attempting such a general listing and classification of the activities of the government upon the completion of the present series.

1. Utilization of National Forest Resources
    1. Formulation of management plans
    2. Timber sale
    3. Control of grazing
    4. Promotion of recreational use
    5. Supervision of the development of private hydroelectric power projects
2. Protection of the National Forests
    1. Fire prevention and fire fighting
    2. Suppression of insect pests
3. Development of the National Forests
    1. Approval of major highway projects
    2. Construction and maintenance of minor roads and trails
    3. Reforestation
4. Research
    1. Range investigations
    2. Silvicultural experiments
    3. Timber and timber products research
    4. Forest taxation studies
    5. Economic and statistical studies
5. Information Service
6. Coöperation with State Governments and with other Services of the National Government.

APPENDIX 3

PUBLICATIONS

The bulk of the publications of the Forest Service consists of scientific and professional papers concerning the work of the Service and the forest problems which it is attempting to solve. A smaller proportion of the Service's published output concerns itself with historical and statistical matter, and literature for the layman, including expositions of the Service and its work.

The scientific and professional papers are issued in various sizes and under a variety of names. Some of them, because of their brevity, may be considered as articles, sketches, reviews, or leaflets. Others are of monograph size. They vary from a few pages to over a hundred, the average being closer to the minimum than the maximum. For example, there may be cited:

Basic Grading Rules and Working Stresses for Structural Timbers. As recommended by the Department of Agriculture By J. A. Newlin, in charge section of timber mechanics, and R. P. A. Johnson, engineer in forest products laboratory, Forest Service. Pp. 23. October, 1923. (Department Circular 295.)

As an example of the paper approaching the monograph class, one may cite:

Natural Reproduction of Western Yellow Pine in the Southwest. By G. A. Pearson, silviculturist, Fort Valley Forest Experiment Station, Forest Service. Pp. 144, figs. 16, pls. 22. April 27, 1923. (Department Bulletin 1105.)

In some cases these papers are published as Farmers' Bulletins and in others as Department Bulletins, Department Circulars, Bureau Bulletins, Bureau Circulars, or Miscellaneous Circulars.[1] Articles of a generally similar nature are published

[1] It is difficult to define with exactness the border lines between these several types of Department of Agriculture publications. Generally speaking, however, Department Bulletins are those publications which embody

in the Journal of Agricultural Research from time to time, to be issued later in leaflet or booklet form under the general designation of Journal of Agricultural Research " Separates." There are Yearbook " Separates " also, which are reprints from the pages of the Yearbook of the Department of Agriculture. The Yearbook publications are usually of a popular nature.

These publications have appeared from time to time, beginning with the earliest years of the Service and, collectively, represent a considerable American forestry literature.

The first Forest Service Bulletin, published in 1887, was a " Report on the Relation of Railroads to Forest Supplies and Forestry." Between that year and 1921, 179 Forest Service and Department Bulletins appeared, the last one in 1921[2] being " Regional Development of the Pulpwood Resources of the Tongass National Forest, Alaska." Forest Service Bulletins selected as typical throughout the period covered, included the following:

No. 5. (1891) What is Forestry? by B. E. Fernow.
No. 24. (1899) A Primer of Forestry, by Gifford Pinchot.
No. 26. (1899) Practical Forestry in the Adirondacks, by Henry Solon Graves.
No. 32. (1902) Working Plan for Forest Lands Near Pine Bluff, Arkansas, by F. E. Olmsted.
No. 33. (1902) Western Hemlock, by E. T. Allen.
No. 97. (1911) Coyote-Proof Inclosures in Connection with Range Lambing Grounds, by J. T. Jardine.
No. 118. (1912) Prolonging the Life of Crossties, by H. F. Weiss.

the results of a piece of scientific investigation. Department Circulars are scientific publications of a preliminary nature. Farmers' Bulletins are presentations in popular form of scientific matter of interest and value to farmers. Miscellaneous Circulars and those publications called " Miscellaneous " contain matter which it is not desired to publish under other designations. The distinction between bureau bulletins and bureau circulars is the same as that between department bulletins and department circulars.

[2] The year 1921 is given because the Forest Service, at the time this text was written, had published lists of its publications up to that year. The " Complete List of Forest Service Publications " covers the period from 1876 to Nov. 15, 1915. The " Supplementary List " takes care of the period intervening between Nov. 15, 1915, and July 31, 1921. A third list bringing the subject up to December, 1925, has since been published. All of these lists, published in mimeograph form, were the work of the Library, Office of Forest Investigations, Branch of Research.

Department Bulletins brought out since 1921 include:

No. 1059. (1922) Research Methods in the Study of Forest Environ-
ment, by C. G. Bates and R. Zon.
No. 1136. (1923) Kiln Drying Handbook, by R. Thelen.
No. 1176. (1923) Some Results of Cutting in the Sierra Forests of
California, by D. Dunning.

There are 216 Forest Service Circulars[3] published between
1886 and 1913; No. 1, in 1886, being "A Request to Educa-
tors for Coöperation." No. 18, " Progress in Timber Physics,"
by B. E. Fernow, appeared in 1898. No. 23, " Suggestions
to Prospective Forest Students," by Gifford Pinchot, was
published in 1902. No. 216, which appeared in 1913, was
" Effect of Forest Fires on Standing Hardwood Timber," by
W. H. Long. Many of these earlier circulars were devoted, each
to a particular tree species, being in this respect similar to a
series of fifty-three " Silvical Leaflets," published between 1907
and 1912. After 1913 no circulars were published until 1919,
in which year " Circulars of the Department of Agriculture "
began to appear, a portion of them being devoted to Forest
Service interests. A considerable proportion of this later series
of circulars has dealt with the recreational features of the national forests, some of them constituting in effect hunting, fishing, and outing guide books. Others are concerned with the
scientific and professional aspects of the subject.

The first Farmers' Bulletin devoted to forestry and prepared
by the then Division of Forestry of the Department of Agriculture appeared in 1898—" Forestry for Farmers," by B. E.
Fernow.[4] Sixteen similar bulletins appeared thereafter up to
1914, one of them being No. 592, "Stock Watering Places," by
W. C. Barnes. Twelve additional bulletins appeared between
1916 and 1921, all but two of them being devoted to some phase

[3] Prior to 1905 they were known as Division or Bureau Circulars.
[4] Previously published in 1894 in the Yearbook of the Department of
Agriculture.

of the farm woodlot question, as the following titles will illustrate:

No. 711. The Care and Improvement of the Woodlot, by C. R. Til-
lotson.
No. 788. The Windbreak as a Farm Asset, by C. G. Bates.
No. 1117. Forestry and Farm Income, by W. R. Mattoon.
No. 1123. Growing and Planting Hardwood Seedlings on the Farm, by
C. R. Tillotson.

Many of the articles prepared by the Forest Service for publication in the Yearbook pertain to farmers' interests. "The Farm Woodlot Problem," by H. A. Smith, appeared in the Yearbook for 1914. "How the Public Forests Are Handled," by the same author, was in the 1920 volume. Yearbook forestry articles that are not primarily of farm interest or application include the following Yearbook Separates:

No. 688. (1916) Farms, Forests, and Erosion, by S. T. Dana.
No. 696. (1916) Opening up the National Forests by Road Building,
by O. C. Merrill.
No. 835. (1920) Wood for the Nation, by W. B. Greeley.

The character of the articles that the Forest Service occasionally publishes in the Journal of Agricultural Research, and not infrequently republishes, as booklets or leaflets, under the general term of "Separates," is different from those published in the Yearbooks. Intended for scientific and professional rather than for popular consumption, they deal altogether with the purely technical phases of forestry; for example, wood structure, peculiarities of species, etc. An article of this sort reprinted in 1921 was "Chlorosis of Conifers Corrected by Spraying with Ferrous Sulphate," by C. F. Korstian.

The June, 1925, issue of the Journal contains an article entitled "Factors Affecting the Reproduction of Engelmann Spruce," by W. C. Lowdermilk.

Under the general designations of Circulars of the Office of the Secretary and Department Reports, there have been put out a number of articles in pamphlet form dealing, in the case of the circulars, usually with propaganda for the more general adoption of forestry methods by private woodland owners, and in the case of the reports with some economic phase of the forestal situation.[5]

The character of the Department reports is sufficiently illustrated by the following titles:

No. 110. (1916) Meat Situation in the United States: pt. 2, Livestock production in eleven far western range states, by W. C. Barnes and J. T. Jardine.

No. 114. (1917) Some Public and Economic Aspects of the Lumber Industry, by W. B. Greeley.

No. 117. (1917) The Substitution of Other Materials for Wood, by R. Thelen.

Space is not available for extended notice of the variety of Forest Service publications that have been brought out since the early days of governmental forestry as separate publications. They are concerned for the most part with utilizational and recreational aspects of the national forests, but not exclusively so. They deal from time to time with the forest development, forest economy, and forest propaganda fields. Many recreation maps and folders for different national forests have been published in this form. The title of a circular printed in 1919, "A Vacation Land of Lakes and Woods: The Superior National Forest," is an illustration. Similarly, there were issued in 1918 three booklets by F. A. Waugh, the titles of which were: " Landscape Engineering in the National Forests," " Recreation Uses on National Forests " and "A Plan for the Development of Village of Grand Canyon, Arizona."

Information concerning the courses given at the Forest Products Laboratory in boxing and crating have likewise been published in the Miscellaneous series; also information concerning national forest homesteads, free tree distribution in Nebraska under the Kinkaid Act, first aid for field parties, and a variety of other subjects.

Among miscellaneous publications may be cited also two published in 1919 and 1920 on the " Purchase of Land for National Forests Under the Act of March 1, 1911," and the

[5] These circulars and reports are no longer published.

"Report on Senate Resolution 311," which, in 1920, dealt with the matter of "Timber Depletion, Lumber Prices, Lumber Exports, and Concentration of Timber Ownership."

A widely used miscellaneous volume is called the "Use Book." It has appeared in a number of editions and revisions since 1905. This manual for the convenience of users or prospective users of the national forests, contains the departmental regulations governing the utilization of the national forest timber, range, recreational, and water power resources, together with directions to the user concerning the officers with whom he must deal, and the rules he must observe. Smaller or subeditions of this work have appeared from time to time covering one forest resource only, such as the grazing edition in 1910 and a water power edition in 1911. The latest edition of the book covering all resources appeared in 1918. Another grazing edition was brought out in 1921.

Publications of the Service which appear annually include the Annual Report of the Forester, a small booklet containing the latest figures on the national forest areas, and a directory of Service personnel.

Beginning September, 1924, the Service published a bimonthly magazine in mimeographed form called "The Forest Worker." This publication has been issued in printed form since January, 1927. It contains news of the Forest Service and of the progress of forestry throughout the country. It is paralleled somewhat by the "Service Bulletin," which is issued weekly in much the same general style as the Forest Worker but with a restriction as to distribution to Forest Service personnel. Its primary purposes are to furnish the personnel with a weekly digest of Service news from all the forest districts, the Madison Laboratory, and Washington headquarters, and to stimulate discussion of live questions of administration, procedure, and organization.

The Forest Service does a considerable amount of map-publishing, almost entirely for the convenience of its own personnel, although occasionally a map is issued for general distribution.

These maps cover the forest regions of America and the forest districts of the Service only.

The publications which have been described above constitute what may be termed the " regular " publications of the Service. They do not exhaust the Service's means of publicity and expression, however. The Service frequently coöperates, either with some other governmental agency, or with some state or private organization, in the preparation of what are known as " Coöperative Reports." The following titles will illustrate the nature and purpose of these reports:

Wood Using Industries in California, 1912. (In coöperation with the California State Board of Forestry.)

Wood Using Industries of Georgia, 1915. (In coöperation with the Lumber Trade Journal.)

Pulpwood Consumption and Wood Pulp Production, 1918. (In cooperation with the Newsprint Service Bureau.)

Glues Used in Airplane Parts, 1920. (In coöperation with the National Advisory Committee for Aeronautics.)

Most of the results of coöperative investigations have been published by the coöperating organizations, but some have appeared as publications of the Forest Service. Thus all of the series on the wood-using industries, two of which are listed above, were published by the coöperating agencies. The report on pulpwood and wood pulp was published by the Forest Service, while the report on glues used in airplane parts was issued by the National Advisory Committee for Aeronautics, a research unit of the national government.

Magazine articles are written by employees of the Forest Service upon various forestal topics. Many such articles by members of the Forest Service appear during the course of a year in professional, trade, and educational journals.

The compilation of information known as the " Atlas " is intended primarily for the use of the Service personnel and is not circulated. It is prepared by the Branch of Engineering. Its pages are open to the public, and comprise an encyclopædia of cartographical, silvicultural, geologic, economic, financial, legal, and other information about every national forest in the country.

APPENDIX 4

LAWS

(A) Index to Laws

11

(B) Compilation of Laws

There are already in existence eight compilations of laws having to do with the American forest. Six of these are government publications; one has been produced privately; one bears a university imprint. The government compilations are:

1. " Compilation of Laws, and Regulations and Decisions Thereunder, Relating to the Creation and Administration of Public Forest Reserves." Issued November 6, 1900, by the General Land Office.

2. " Compilation of Public Timber Laws, and Regulations and Decisions Thereunder." A General Land Office publication, issued February 14, 1903.

3. " Compilation of Laws, and Regulations and Decisions Thereunder, Relating to the Establishment of Federal Forest Reserves under Section 24 of the Act of March 3, 1891 (26 Stat. L., 1095), and the Administration Thereof." This was also a General Land Office publication. It was approved by the Secretary of the Interior, October 3, 1903, and was issued, presumably, shortly thereafter.

4. " Federal and State Forest Laws," Compiled by George W. Woodruff, Expert, Bureau of Forestry. This was a Department of Agriculture publication—Bureau of Forestry Bulletin, No. 57. Issued December, 1904.

5. " Laws, Decisions, and Opinions Applicable to the National Forests." Revised and Compiled by R. F. Feagans under the direction of the Solicitor of the Department of Agriculture, 1916.

6. " Laws Applicable to the United States Department of Agriculture, 1923." An Agricultural Department publication, " printed for the use of officers and employees of the Department." Issued 1924.[1] Pages 285-474 refer to the Forest Service.

The two compilations which were not issued by the government are both the work of Mr. J. P. Kinney, of the Forestry Branch of the Indian Service. One, " Forest Legislation in America Prior to March 4, 1789," was published by Cornell University in January, 1916, as Bulletin 370. It was part of a study presented in partial fulfillment of the requirements for a degree in forestry. The other, " The Development of Forest Law in America," was published in 1917, by John Wiley and Sons, Inc.

Only general laws applicable to several forests are given in this compilation. In addition, there is an appreciable number of acts relating to specific forests or areas.

1876—Act of August 15, 1876 (19 Stat. L., 143, 167)—An Act Making appropriations for the legislative, executive and judicial expenses of the Government for the year ending June thirtieth, eighteen hundred and seventy-seven, and for other purposes.

* * * *

For purchase and distribution of new and valuable seeds and plants, sixty thousand dollars: *Provided,* That two thousand dollars of the above amount shall be expended by the Commissioner of Agriculture as compensation to some man of approved attainments, who is practically well acquainted with methods of statistical inquery, and who has evinced an intimate acquaintance with questions relating to the national wants in regard to timber to prosecute investigations and inqueries, with the view of ascertaining the annual amount of consumption, importation, and exportation of timber and other forest-products, the probable supply for future wants, the means best adapted to their

---

[1] The first edition of this compilation was issued in 1908. Pages 131-64 refer to the Forest Service. A revision of this work was brought out in 1912, pp. 86-183 of which refer to the Forest Service. Three supplements to this revision were published within the next three years. Of the first, issued in the latter part of 1913, pp. 22-36 are devoted to the Forest Service. Of the second and third supplements, issued on Feb. 19 and July 6, 1915, respectively, pp. 31-68 of the one, and 25-38 of the other, are so devoted.

preservation and renewal, the influence of forests upon climate, and the measures that have been successfully applied in foreign countries, or that may be deemed applicable in this country, for the preservation and restoration or planting of forests; and to report upon the same to the Commissioner of Agriculture to be by him in a separate report transmitted to Congress. . . . .

1897—Act of June 4, 1897 (30 Stat. L., 11, 34, 35, 36)—An Act Making appropriations for sundry civil expenses of the Government for the fiscal year ending June thirtieth, eighteen hundred and ninety-eight, and for other purposes.

* * * *

All public lands heretofore designated and reserved by the President of the United States under the provisions of the Act approved March third, eighteen hundred and ninety-one, the orders for which shall be and remain in full force and effect, unsuspended and unrevoked, and all public lands that may hereafter be set aside and reserved as public forest reserves under said Act, shall be as far as practicable controlled and administered in accordance with the following provisions:

No public forest reservation shall be established except to improve and protect the forest within the reservation, or for the purpose of securing favorable conditions of water flows, and to furnish a continuous supply of timber for the use and necessities of citizens of the United States; but it is not the purpose or intent of these provisions, or of the Act providing for such reservations, to authorize the inclusion therein of lands more valuable for the mineral therein, or for agricultural purposes, than for forest purposes.

The Secretary of the Interior shall make provisions for the protection against destruction by fire and depredations upon the public forests and forest reservations which may have been set aside or which may be hereafter set aside under the said Act of March third, eighteen hundred and ninety-one, and which may be continued; and he may make such rules and regulations and establish such service as will insure the objects of such reservations, namely, to regulate their occupancy and use and to preserve the forests thereon from destruction; and any violation of the provisions of this Act or such rules and regulations shall be punished as is provided for in the Act of June fourth, eighteen hundred and eighty-eight, amending section fifty-three hundred and eighty-eight of the Revised Statutes of the United States.

For the purpose of preserving the living and growing timber and promoting the younger growth on forest reservations, the Secretary of the Interior, under such rules and regulations as he shall prescribe, may cause to be designated and appraised so much of the dead, matured, or large growth of trees found upon such forest reservations as may be compatible with the utilization of the forests thereon, and may sell the same for not less than the appraised value in such quantities to each purchaser as he shall prescribe, to be used in the State or Territory in which such timber reservation may be situated, respectively, but not for export therefrom. [As amended June 6, 1900 (31 Stat. L., 661).] Before such sale shall take place notice thereof shall be given by the Commissioner of the General Land Office, for not less than thirty days, by publication in one or more newspapers of general circulation, as he may deem necessary ,in the State or Territory where such reservation exists: *Provided, however,* That in cases of unusual emergency the Secretary of the Interior may, in the exercise of his discretion, permit the purchase of timber and cord wood in advance of advertisement of sale at rates of value approved by him and subject to payment of the full amount of the highest bid resulting from the usual advertisement of sale: *Provided further,* That he may, in his discretion, sell without advertisement, in quantities to suit applicants, at a fair appraisement, timber and cord wood not exceeding in value one hundred dollars stumpage: [2] *And provided further,* That in cases in which advertisement is had and no satisfactory bid is received, or in cases in which the bidder fails to complete the purchase, the timber may be sold, without further advertisement, at private sale, in the discretion of the Secretary of the Interior, at not less than the appraised valuation, in quantities to suit purchasers: [The original act continues] payments for such timber to be made to the receiver of the local land office of the district wherein said timber may be sold, under such rules and regulations as the Secretary of the Interior may prescribe; and the moneys arising therefrom shall be accounted for by the receiver of such land office to the Commissioner of the General Land Office, in a separate account, and shall be covered into the Treasury. Such timber. before being sold, shall be marked and designated, and shall be cut and removed under the supervision of some person appointed for that purpose by the Secretary of the Interior, not interested in the purchase or removal of such timber nor in the employment of the purchaser thereof. Such supervisor shall make report in writing to the Commissioner of the General Land Office and to the receiver in the land office in which such reservation shall be located of his doings in the premises.

The Secretary of the Interior may permit, under regulations to be prescribed by him, the use of timber and stone found upon such reservations, free of charge, by bona fide settlers, miners, residents, and prospectors for minerals, for firewood, fencing, buildings, mining, prospecting, and other domestic purposes, as may be needed by such persons for such purposes; such timber to be used within the State or Territory, respectively, where such reservations may be located.

Nothing herein shall be construed as prohibiting the egress or ingress of actual settlers residing within the boundaries of such reservations,

[2] Amended by Section 3 of act of March 3, 1925 (43 Stat. L., 1132), which raised the value limit to $500.

or from crossing the same to and from their property or homes; and such wagon roads and other improvements may be constructed thereon as may be necessary to reach their homes and to utilize their property under such rules and regulations as may be prescribed by the Secretary of the Interior. Nor shall anything herein prohibit any person from entering upon such forest reservations for all proper and lawful purposes, including that of prospecting, locating, and developing the mineral resources thereof: *Provided,* That such persons comply with the rules and regulations covering such forest reservations.

* * * *

The settlers residing within the exterior boundaries of such forest reservations, or in the vicinity thereof, may maintain schools and churches within such reservation, and for that purpose may occupy any part of the said forest reservation, not exceeding two acres for each school house and one acre for a church.

The jurisdiction, both civil and criminal, over persons within such reservations shall not be affected or changed by reason of the existence of such reservations, except so far as the punishment of offenses against the United States therein is concerned; the intent and meaning of this provision being that the State wherein any such reservation is situated shall not, by reason of the establishment thereof, lose its jurisdiction nor the inhabitants thereof their rights and privileges as citizens, or be absolved from their duties as citizens of the State.

All waters on such reservations may be used for domestic, mining, milling, or irrigation purposes, under the laws of the State wherein such forest reservations are situated, or under the laws of the United States and the rules and regulations established thereunder.

Upon the recommendation of the Secretary of the Interior, with the approval of the President, after sixty days' notice thereof, published in two papers of general circulation in the State or Territory wherein any forest reservation is situated, and near the said reservation, any public lands embraced within the limits of any forest reservation which, after due examination by personal inspection of a competent person appointed for that purpose by the Secretary of the Interior, shall be found better adapted for mining or for agricultural purposes than for forest usage, may be restored to the public domain. And any mineral lands in any forest reservation which have been or which may be shown to be such, and subject to entry under the existing mining laws of the United States and the rules and regulations applying thereto, shall continue to be subject to such location and entry, notwithstanding any provisions herein contained.

The President is hereby authorized at any time to modify any Executive order that has been or may hereafter be made establishing any forest reserve, and by such modification may reduce the area or change the boundary lines of such reserve, or may vacate altogether any order creating such reserve.

1899—Act of February 28, 1899 (30 Stat. L., 908)—An Act
To authorize the Secretary of the Interior to rent or
lease certain portions of forest reserve.

That the Secretary of the Interior be, and hereby is, authorized,
under such rules and regulations as he from time to time may make, to
rent or lease to responsible persons or corporations applying therefor
suitable spaces and portions of ground near, or adjacent to, mineral,
medicinal, or other springs, within any forest reserves established within
the United States, or hereafter to be established, and where the public
is accustomed or desires to frequent, for health or pleasure, for the
purpose of erecting upon such leased ground sanitariums or hotels, to
be opened for the reception of the public. And he is further authorized
to make such regulations, for the convenience of people visiting such
springs, with reference to spaces and locations, for the erection of tents
or temporary dwelling houses to be erected or constructed for the use of
those visiting such springs for health or pleasure. And the Secretary
of the Interior is authorized to prescribe the terms and duration and
the compensation to be paid for the privileges granted under the provisions of this Act.[3]

1899—Act of March 3, 1899 (30 Stat. L., 1074, 1095)—An Act
Making appropriations for sundry civil expenses of
the Government for the fiscal year ending June thirtieth, nineteen hundred, and for other purposes.

* * * *

Protection and Administration of Forest Reserves: . . . *Provided
further,* That forest agents, superintendents, supervisors, and all other
persons employed in connection with the administration and protection of forest reservations shall in all ways that are practicable, aid in
the enforcement of the laws of the State or Territory in which said
forest reservation is situated, in relation to the protection of fish and
game. . . .

[3] The powers herein granted are stated in broader terms in Act of March
4, 1915 (38 Stat. L., 1086, 1101) as follows:
" That hereafter the Secretary of Agriculture may, upon such terms as
he may deem proper, for periods not exceeding thirty years, permit responsible persons or associations to use and occupy suitable spaces or portions
of ground in the national forests for the construction of summer homes,
hotels, stores, or other structures needed for recreation or public convenience,
not exceeding five acres to any one person or association, but this shall not
be construed to interfere with the right to enter homesteads upon agricultural lands in national forests as now provided by law."

1899—Act of March 3, 1899 (30 Stat. L., 1214, 1233)—An Act Making appropriations to supply deficiencies in the appropriations for the fiscal year ending June thirtieth, eighteen hundred and ninety-nine, and for prior years, and for other purposes.

* * * *

That in the form provided by existing law the Secretary of the Interior may file and approve surveys and plats of any right of way for a wagon road, railroad, or other highway over and across any forest reservation or reservoir site when in his judgment the public interests will not be injuriously affected thereby.

1905—Act of January 24, 1905 (33 Stat. L., 614)—An Act For the protection of wild animals and birds in the Wichita Forest Reserve.

[SECTION 1]. That the President of the United States is hereby authorized to designate such areas in the Wichita Forest Reserve as should, in his opinion, be set aside for the protection of game animals and birds and be recognized as a breeding place therefor.

SEC. 2. That when such areas have been designated as provided for in section one of this Act, hunting, trapping, killing, or capturing of game animals and birds upon the lands of the United States within the limits of said areas shall be unlawful, except under such regulations as may be prescribed from time to time, by the Secretary of Agriculture; and any person violating such regulations or the provisions of this Act shall be deemed guilty of a misdemeanor, and shall upon conviction in any United States court of competent jurisdiction, be fined in a sum not exceeding one thousand dollars or be imprisoned for a period not exceeding one year, or shall suffer both fine and imprisonment, in the discretion of the court.

SEC. 3. That it is the purpose of this Act to protect from trespass the public lands of the United States and the game animals and birds which may be thereon, and not to interfere with the operation of the local game laws as affecting private, State or Territorial lands.[4]

⁴ The preserve was formally created by Presidential Proclamation on June 2, 1905 (34 Stat. L., 3062). A clause in the act of June 30, 1906 (34 Stat. L., 669, 696), authorized and made appropriation for sheds and fences on the new preserve. In the same year another game preserve was created in the Grand Canyon National Forest by the same method—act of June 29, 1906 (34 Stat. L., 607), and proclamation of November 28, 1906 (34 Stat. L., 3263). A clause in the act of August 11, 1916 (39 Stat. L., 446, 476), authorized the creation of similar preserves in forests acquired under the Weeks Act and under that authority the Pisgah Preserve was set aside by procla-

1905—Act of February 1, 1905 (33 Stat. L., 628)—An Act
Providing for the transfer of forest reserves from the
Department of the Interior to the Department of Agriculture.

[SECTION 1]. That the Secretary of the Department of Agriculture
shall, from and after the passage of this Act, execute or cause to be
executed all laws affecting public lands heretofore or hereafter reserved
under the provisions of section twenty-four of the Act entitled "An Act
to repeal the timber-culture laws, and for other purposes," approved
March third, eighteen hundred and ninety-one, and Acts supplemental
to and amendatory thereof, after such lands have been so reserved,
excepting such laws as affect the surveying, prospecting, locating, appropriating, entering, relinquishing, reconveying, certifying, or patenting
of any of such lands.

1905—Act of February 8, 1905 (33 Stat. L., 706)—An Act
Authorizing the use of earth, stone and timber on the
public lands and forest reserves of the United States
in the construction of works under the national irrigation law.

[SECTION 1]. That in carrying out the provisions of the national
irrigation law, approved June seventeenth, nineteen hundred and two,
and in constructing works thereunder . . . the Secretary of Agriculture
is hereby authorized to permit the use of earth, stone, and timber from
the forest reserves of the United States . . . under rules and regulations
to be prescribed by him.

---

mation of October 17, 1916 (39 Stat. L., 1811), and the Cherokee Refuges
Numbers One and Two were designated and set aside by proclamation
of August 5, 1924 (43 Stat. L., 1964). By the act of June 5, 1920 (41 Stat. L.,
986), and the Proclamation of October 9, 1920 (41 Stat. L., 1805), the
Custer State Park Game Sanctuary in South Dakota was similarly created.
The Act of June 7, 1924 (43 Stat. L., 594) authorized the designation of a
game refuge within the Medicine Bow National Forest, and the Proclamation of August 8, 1924 (43 Stat. L., 1964) formally created the game refuge
within that Forest known as Sheep Mountain. The Act of February 28,
1925 (43 Stat. L., 1091) authorized the creation of game refuges on the
Ozark National Forest in the State of Arkansas, and Ozark Refuges Numbers One, Two, Three, and Four, were created by Presidential Proclamation
on April 26, 1926 (44 Stat. L., 2611). The Sequoia National Game Refuge
was definitely created by Act of July 3, 1926 (44 Stat. L., 821), no proclamation being required to make the act effective. The Act of June 28, 1930
(Public No. 466, 71st Congress) authorized the creation of game sanctuaries
or refuges within the Ocala National Forest in Florida but the land to be set
aside has not yet been designated.

1906—Act of June 8, 1906 (34 Stat. L., 225)—An Act For the preservation of American antiquities.

That any person who shall appropriate, excavate, injure, or destroy any historic or prehistoric ruin or monument, or any object of antiquity, situated on lands owned or controlled by the Government of the United States, without the permission of the Secretary of the Department of the Government having jurisdiction over the lands on which said antiquities are situated, shall, upon conviction, be fined in a sum of not more than five hundred dollars or be imprisoned for a period of not more than ninety days, or shall suffer both fine and imprisonment, in the discretion of the court.

SEC. 2. That the President of the United States is hereby authorized, in his discretion, to declare by public proclamation historic landmarks, historic and prehistoric structures, and other objects of historic or scientific interest that are situated upon the lands owned or controlled by the Government of the United States to be national monuments, and may reserve as a part thereof parcels of land, the limits of which in all cases shall be confined to the smallest area compatible with the proper care and management of the objects to be protected: *Provided*, That when such objects are situated upon a tract covered by a bona fide unperfected claim or held in private ownership, the tract, or so much thereof as may be necessary for the proper care and management of the object, may be relinquished to the Government, and the Secretary of the Interior is hereby authorized to accept the relinquishment of such tracts in behalf of the Government of the United States.

SEC. 3. That permits for the examination of ruins, the excavation of archæological sites, and the gathering of objects of antiquity upon the lands under their respective jurisdictions may be granted by the Secretaries of the Interior, Agriculture, and War to institutions which they may deem properly qualified to conduct such examination, excavation, or gathering, subject to such rules and regulations as they may prescribe; *Provided*, That the examinations, excavations, and gatherings are undertaken for the benefit of reputable museums, universities, colleges, or other recognized scientific or educational institutions, with a view to increasing the knowledge of such objects, and that the gatherings shall be made for permanent preservation in public museums.

SEC. 4. That the Secretaries of the Departments aforesaid shall make and publish from time to time uniform rules and regulations for the purpose of carrying out the provisions of this act.

1906—Act of June 11, 1906 (34 Stat. L., 233)—An Act To provide for the entry of Agricultural lands within forest reserves.

[SECTION 1]. That the Secretary of Agriculture may, in his discretion, and he is hereby authorized, upon application or otherwise, to examine and ascertain as to the location ·and extent of lands within permanent or temporary forest reserves. . . .⁵ which are chiefly valuable for agriculture, and which, in his opinion, may be occupied for agricultural purposes without injury to the forest reserves, and which are not needed for public purposes, and may list and describe the same by metes and bounds, or otherwise, and file the lists and descriptions with the Secretary of the Interior, with the request that the said lands be opened to entry in accordance with the provisions of the homestead laws and this Act.

Upon the filing of any such list or description the Secretary of the Interior shall declare the said lands open to homestead settlement and entry in tracts not exceeding one hundred and sixty acres in area and not exceeding one mile in length, at the expiration of sixty days from the filing of the list in the land office of the district within which the lands are located, during which period the said list or description shall be prominently posted in the land office and advertised for a period of not less than four weeks in one newspaper of general circulation published in the county in which the lands are situated: *Provided,* That any settler actually occupying and in good faith claiming such lands for agricultural purposes prior to January first, nineteen hundred and six, and who shall not have abandoned the same, and the person, if qualified to make a homestead entry, upon whose application the land proposed to be entered was examined and listed, shall, each in the order named, have a preference right of settlement and entry: *Provided further,* That any entryman desiring to obtain patent to any lands described by metes and bounds entered by him under the provisions of this Act shall, within five years of the date of making settlement, file, with the required proof of residence and cultivation, a plat and field notes of the lands entered, made by or under the direction of the United States surveyor-general, showing accurately the boundaries of such lands, which shall be distinctly marked by monuments on the ground, and by posting a copy of such plat, together with a notice of the time and place of offering proof, in a conspicuous place on the land embraced in such plat during the period prescribed by law for the publication of his notice of intention to offer proof, and that a copy of such plat and field notes shall also be kept posted in the office of the register of the land office for the land district in which such lands are situated for a like period; and further, that any agricultural lands within forest reserves may, at the discretion of the Secretary, be surveyed by metes and bounds, and that no lands entered under the provisions of this Act shall be patented under the commutation provisions of the homestead laws, but settlers, upon final proof, shall have credit for the period of their actual residence upon the lands covered by their entries.

⁵ Certain counties in California were excluded from the provisions of the original act, but the section was amended by the act of May 30, 1908 (35 Stat. L., 554), to make the act applicable to all counties in California except those of San Luis Obispo and Santa Barbara.

Sec. 2. That settlers upon lands chiefly valuable for agriculture within forest reserves on January first, nineteen hundred and six, who have already exercised or lost their homestead privilege, but are otherwise competent to enter lands under the homestead laws, are hereby granted an additional homestead right of entry for the purposes of this Act only, and such settlers must otherwise comply with the provisions of the homestead law, and in addition thereto must pay two dollars and fifty cents per acre for lands entered under the provisions of this section, such payment to be made at the time of making final proof on such lands.

Sec. 3. That all entries under this Act in the Black Hills Forest Reserve shall be subject to the quartz or lode mining laws of the United States, and the laws and regulations permitting the location, appropriation, and use of the waters within the said forest reserves for mining, irrigation, and other purposes; and no titles acquired to agricultural lands in said Black Hills Forest Reserve under this Act shall vest in the patentee any riparian rights to any stream or streams of flowing water within said reserve; and that such limitation of title shall be expressed in the patents for the lands covered by such entries.

Sec. 4. [This section excluded from the operation of the act portions of the Black Hills Forest Reserve in Lawrence and Pennington counties in South Dakota. This exclusion was repealed by act of Aug. 8, 1916 (39 Stat. L., 440).]

Sec. 5. That nothing herein contained shall be held to authorize any future settlement on any lands within forest reserves until such lands have been opened to settlement as provided in this Act, or to in any way impair the legal rights of any bona fide homestead settler who has or shall establish residence upon public lands prior to their inclusion within a forest reserve.

1907—Act of March 4, 1907 (34 Stat. L., 1256, 1269, 1270)—
An Act Making appropriations for the Department of Agriculture for the fiscal year ending June thirtieth, nineteen hundred and eight.

*   *   *   *

. . . forest reserves . . . shall be known hereafter as national forests.
*   *   *   *

That all money received after July first, nineteen hundred and seven, by or on account of the forest service for timber, or from any other source of forest reservation revenue, shall be covered into the Treasury of the United States as a miscellaneous receipt and there is hereby appropriated and made available as the Secretary of Agriculture may direct out of any funds in the Treasury not otherwise appropriated, so much as may be necessary to make refunds to depositors of money heretofore or hereafter deposited by them to secure the purchase price on the sale of any products or for the use of any land or resources of the national forests in excess of amounts found actually due from them to the United States.[6]

1908—Act of May 23, 1908 (35 Stat. L., 251, 259)—An Act Making appropriations for the Department of Agriculture for the fiscal year ending June thirtieth, nineteen hundred and nine.

* * * *

No part of this appropriation shall be used for any experiment or test made outside the jurisdiction of the United States.[7]

* * * *

Hereafter officials of the Forest Service . . . with respect to National Forests, shall aid the other Federal Bureaus and Departments on request from them, in the performance of the duties imposed on them by law.

* * * *

Hereafter advances of money under any appropriation for the Forest Service may be made to the Forest Service and by authority of the Secretary of Agriculture to chiefs of field parties for fighting forest fires in emergency cases, who shall give bond under such rules and regulations and in such sum as the Secretary of Agriculture may direct, and detailed accounts arising under such advances shall be rendered through and by the Department of Agriculture to the Treasury Department.

* * * *

And there is hereby appropriated out of any money in the Treasury not otherwise appropriated, the sum of six hundred thousand dollars to be expended as the Secretary of Agriculture may direct, for the construction and maintenance of roads, trails, bridges, fire lanes, telephone lines, cabins, fences, and other permanent improvements necessary for the proper and economical administration, protection, and development of the National Forests.[8]

* * * *

That hereafter twenty-five per centum of all money received from each forest reserve during any fiscal year, including the year ending June thirtieth, nineteen hundred and eight, shall be paid at the end thereof by the Secretary of the Treasury to the State or Territory in

[6] Amended by Act of March 4, 1911 (36 Stat. L., 1235, 1253).

[7] Repeated in every subsequent appropriation act.

[8] A part of every subsequent appropriation act. Since 1918 (act of March 4, 1917) the act has carried a proviso limiting expenditures for fences, corrals, and certain other specific objects. In 1909, 1910, and 1911 the appropriations were under a special head, " Improvement of the National Forests." Since 1911 under " General Expenses."

which said reserve is situated, to be expended as the State or Territorial legislature may prescribe for the benefit of the public schools and public roads of the county or counties in which the forest reserve is situated: *Provided,* That when any forest reserve is in more than one State or Territory or county the distributive share to each from the proceeds of said reserve shall be proportional to its area therein.[9]

1910—Act of June 20, 1910 (36 Stat. L., 557, 561)—An Act To enable the people of New Mexico to form a constitution and state government and be admitted into the Union on an equal footing with the original States; and to enable the people of Arizona to form a constitution and state government and be admitted into the Union on an equal footing with the original States.

* * * *

Sec. 6. That in addition to sections sixteen and thirty-six, heretofore granted to the Territory of New Mexico, sections two and thirty-two in every township in said proposed State not otherwise appropriated at the date of the passage of this Act are hereby granted to the said State for the support of common schools; . . . *And provided further,* That the grants of sections two, sixteen, thirty-two and thirty-six to said State, within national forests now existing or proclaimed, shall not vest the title to said sections in said State until the part of said national forests embracing any of said sections is restored to the public domain; but said granted sections shall be administered as a part of said forests, and at the close of each fiscal year there shall be paid by the Secretary of the Treasury to the State, as income for its common-school fund, such proportion of the gross proceeds of all the national forests within said State as the area of lands hereby granted to said State for school purposes which are situated within said forest reserves, whether surveyed or unsurveyed, and for which no indemnity has been selected, may bear to the total area of all the national forests within said State, the area of said sections when unsurveyed to be determined by the Secretary of the Interior, by protraction or otherwise, the amount necessary for such payments being appropriated and made available annually from any money in the Treasury not otherwise appropriated.[10]

[9] A provision to a similar effect was made in the act making appropriations for the fiscal year 1907, but including the year 1906 (act of June 30, 1906, 34 Stat. L., 669, 684), the amount being, however, 10 per cent. A like provision is found in the following year (34 Stat. L., 1256, 1270).

[10] A similar grant, in identical wording, is made to Arizona in Section 24, pp. 572, 573.

1910—Act of June 25, 1910 (36 Stat. L., 855, 863)—An Act
To provide for determining the heirs of deceased Indians, for the disposition and sale of allotments of deceased Indians, for the leasing of allotments, and for other purposes.

* * * *

SEC 31. That the Secretary of the Interior is hereby authorized, in his discretion, to make allotments within the national forests in conformity with the general allotment laws as amended by section   of this Act, to any Indian occupying, living on, or having improvements on land included within any such national forest who is not entitled to an allotment on any existing Indian reservation, or for whose tribe no reservation has been provided, or whose reservation was not sufficient to afford an allotment to each member thereof. All applications for allotments under the provisions of this section shall be submitted to the Secretary of Agriculture, who shall determine whether the lands applied for are more valuable for agricultural or grazing purposes than for the timber found thereon; and if it be found that the lands applied for are more valuable for agricultural or grazing purposes, then the Secretary of the Interior shall cause allotment to be made as herein provided.[11]

1911—Act of March 1, 1911 (36 Stat. L., 961)—An Act To enable any State to coöperate with any other State or States, or with the United States, for the protection of the watersheds of navigable streams, and to appoint a commission for the acquisition of lands for the purpose of conserving the navigability of navigable rivers.[12]

[SECTION 1]. That the consent of the Congress of the United States is hereby given to each of the several States of the Union to enter into any agreement or compact, not in conflict with any law of the United States, with any other State or States for the purpose of conserving the forests and the water supply of the States entering into such agreement or compact.

SEC. 2. That the sum of two hundred thousand dollars is hereby appropriated and made available until expended, out of any moneys in the National Treasury not otherwise appropriated, to enable the Secretary of Agriculture to coöperate with any State or group of States, when requested to do so, in the protection from fire of the forested water-

[11] Does not apply to Minnesota National Forest.
[12] The Weeks Act.

sheds of navigable streams; and the Secretary of Agriculture is hereby authorized, and on such conditions as he deems wise, to stipulate and agree with any State or group of States to coöperate in the organization and maintenance of a system of fire protection on any private or state forest lands within such State or States and situated upon the watershed of a navigable river: *Provided*, That no such stipulation or agreement shall be made with any State which has not provided by law for a system of forest-fire protection: *Provided further*, That in no case shall the amount expended in any State exceed in any fiscal year the amount appropriated by that State for the same purpose during the same fiscal year.

SEC. 3. That there is hereby appropriated, for the fiscal year ending June thirtieth, nineteen hundred and ten, the sum of one million dollars, and for each fiscal year thereafter a sum not to exceed two million dollars for use in the examination, survey, and acquirement of lands located on the headwaters of navigable streams or those which are being or which may be developed for navigable purposes: *Provided*, That the provisions of this section shall expire by limitation on the thirtieth day of June, nineteen hundred and fifteen.

SEC. 4. That a commission to be known as the National Forest Reservation Commission, consisting of the Secretary of War, the Secretary of the Interior, the Secretary of Agriculture, and two members of the Senate, to be selected by the President of the Senate, and two members of the House of Representatives, to be selected by the Speaker, is hereby created and authorized to consider and pass upon such lands as may be recommended for purchase as provided in section six of this Act, and to fix the price or prices at which such lands may be purchased, and no purchase shall be made of any lands until such lands have been duly approved for purchase by said commission: *Provided*, That the members of the commission herein created shall serve as such only during their incumbency in their respective official positions, and any vacancy on the commission shall be filled in the manner as the original appointment.

SEC. 5. That the commission hereby appointed shall, through its president, annually report to Congress, not later than the first Monday in December, the operations and expenditures of the commission, in detail, during the preceding fiscal year.

SEC. 6.[13] That the Secretary of Agriculture is hereby authorized and directed to examine, locate, and recommend for purchase such lands as in his judgment may be necessary to the regulation of the flow of navigable streams, and to report to the National Forest Reservation Commission the results of such examinations: *Provided*, That before any lands are purchased by the National Forest Reservation Commission said lands shall be examined by the Geological Survey and a report made to the Secretary of Agriculture, showing that the control

[13] Amended by Section 6 of the Clarke-McNary Act of June 7, 1924 (43 Stat. L., 653).

of such lands will promote or protect the navigation of streams on whose watersheds they lie.

SEC. 7. That the Secretary of Agriculture is hereby authorized to purchase, in the name of the United States, such lands as have been approved for purchase by the National Forest Reservation Commission at the price or prices fixed by said commission: *Provided,* That no deed or other instrument of conveyance shall be accepted or approved by the Secretary of Agriculture under this Act until the legislature of the State in which the land lies shall have consented to the acquisition of such land by the United States for the purpose of preserving the navigability of navigable streams. [Section 7 was amended by an act of March 3, 1925 (43 Stat. L., 1215) which added further provisos as follows]:

"*Provided further,* That with the approval of the National Forest Reservation Commission as provided by sections 6 and 7 of this Act, and when the public interests will be benefited thereby, the Secretary of Agriculture be, and hereby is, authorized, in his discretion, to accept on behalf of the United States title to any lands within the exterior boundaries of national forests acquired under this Act which, in his opinion, are chiefly valuable for the purposes of this Act, and in exchange therefor to convey by deed not to exceed an equal value of such national forest land in the same State, or he may authorize the grantor to cut and remove an equal value of timber within such national forests in the same State, the values in each case to be determined by him: *And provided further,* That before any such exchange is effected notice of the contemplated exchange reciting the lands involved shall be published once each week for four successive weeks in some newspaper of general circulation in the county or counties in which may be situated the lands to be accepted, and in some like newspaper published in any county in which may be situated any lands or timber to be given in such exchange. Timber given in such exchanges shall be cut and removed under the laws and regulations relating to such national forests, and under the direction and supervision and in accordance with the requirements of the Secretary of Agriculture. Lands so accepted by the Secretary of Agriculture shall, upon acceptance, become parts of the national forests within whose exterior boundaries they are located, and be subject to all the provisions of this Act.

SEC. 8. That the Secretary of Agriculture may do all things necessary to secure the safe title in the United States to the lands to be acquired under this Act, but no payment shall be made for any such lands until the title shall be satisfactory to the Attorney-General and shall be vested in the United States.

SEC. 9. [As amended by act of March 4, 1913 (37 Stat. L., 828, 855)]:

That such acquisition by the United States shall in no case be defeated because of located or defined rights of way, easements, and

12

reservations, which, from their nature will, in the opinion of the National Forest Reservation Commission and the Secretary of Agriculture, in no manner interfere with the use of the lands so encumbered, for the purposes of the Act: *Provided,* That such rights of way, easements, and reservations retained by the owner from whom the United States receives title, shall be subject to the rules and regulations prescribed by the Secretary of Agriculture for their occupation, use, operation, protection, and administration, and that such rules and regulations shall be expressed in and made a part of the written instrument conveying title to the lands to the United States; and the use, occupation, and operation of such rights of way, easements, and reservations shall be under, subject to, and in obedience with the rules and regulations so expressed.

Sec. 10. That inasmuch as small areas of land chiefly valuable for agriculture may of necessity or by inadvertence be included in tracts acquired under this Act, the Secretary of Agriculture may, in his discretion, and he is hereby authorized, upon application or otherwise, to examine and ascertain the location and extent of such areas as in his opinion may be occupied for agricultural purposes without injury to the forests or to stream flow and which are not needed for public purposes, and may list and describe the same by metes and bounds, or otherwise, and offer them for sale as homesteads at their true value, to be fixed by him, to actual settlers, in tracts not exceeding eighty acres in area, under such joint rules and regulations as the Secretary of Agriculture and the Secretary of the Interior may prescribe; and in case of such sale the jurisdiction over the lands sold shall, *ipso facto,* revert to the State in which the lands sold lie. And no right, title, interest, or claim in or to any lands acquired under this Act, or the waters thereon, or the products, resources, or use thereof after such lands shall have been so acquired, shall be initiated or perfected, except as in this section provided.

Sec. 11. That, subject to the provisions of the last preceding section, the lands acquired under this Act shall be permanently reserved, held, and administered as national forest lands under the provisions of section twenty-four of the Act approved March third, eighteen hundred and ninety-one (volume twenty-six, Statutes at Large, page eleven hundred and three), and Acts supplemental to and amendatory thereof. And the Secretary of Agriculture may from time to time divide the lands acquired under this Act into such specific national forests and so designate the same as he may deem best for administrative purposes.

Sec. 12. That the jurisdiction, both civil and criminal, over persons upon the lands acquired under this Act shall not be affected or changed by their permanent reservation and administration as national forest lands, except so far as the punishment of offenses against the United States is concerned, the intent and meaning of this section being that the State wherein such land is situated shall not, by reason of such reservation and administration, lose its jurisdiction nor the inhabitants thereof their rights and privileges as citizens or be absolved from their duties as citizens of the State.

SEC. 13. That twenty-five [14] per centum of all moneys received during any fiscal year from each national forest into which the lands acquired under this Act may from time to time be divided shall be paid, at the end of such year, by the Secretary of the Treasury to the State in which such national forest is situated, to be expended as the state legislature may prescribe for the benefit of the public schools and public roads of the county or counties in which such national forest is situated: *Provided,* That when any national forest is in more than one State or county the distributive share to each from the proceeds of such forests shall be proportional to its area therein: *Provided further,* That there shall not be paid to any State for any county an amount equal to more than forty per centum of the total income of such county from all other sources.

SEC. 14. That a sum sufficient to pay the necessary expenses of the commission and its members, not to exceed an annual expenditure of twenty-five thousand dollars, is hereby appropriated out of any money in the Treasury not otherwise appropriated. Said appropriation shall be immediately available, and shall be paid out on the audit and order of the president of the said commission, which audit and order shall be conclusive and binding upon all departments as to the correctness of the accounts of said commission.[15]

1911—Act of March 4, 1911 (36 Stat. L., 1235, 1253, 1254)— An Act Making appropriations for the Department of Agriculture for the fiscal year ending June thirtieth, nineteen hundred and twelve.

* * * *

Provided, further, That so much of an Act entitled "An Act making appropriations for the Department of Agriculture for the fiscal year ending June thirtieth, nineteen hundred and eight," approved March fourth, nineteen hundred and seven (Thirty-fourth Statutes at Large, pages twelve hundred and fifty-six and twelve hundred and seventy), which provides for refunds by the Secretary of Agriculture to depositors of moneys to secure the purchase price of timber or the use of lands or resources of the national forests such sums as may be found to be in excess of the amounts found actually due the United States, be, and is

[14] As amended by act of June 30, 1914 (38 Stat. L., 415, 441). The original act had " five " per centum.

[15] See clauses in the act of August 11, 1916 (39 Stat. L., 446, 462, 476), relating to mineral development and game protection on Weeks Law Forests.

hereby, amended hereafter to appropriate and to include so much as may be necessary to refund or pay over to the rightful claimants such sums as may be found by the Secretary of Agriculture to have been erroneously collected for the use of any lands, or for timber or other resources sold from lands located within, but not a part of, the national forests, or for alleged illegal acts done upon such lands, which acts are subsequently found to have been proper and legal; and the Secretary of Agriculture shall make annual report to Congress of the amounts refunded hereunder.

* * * *

That the head of the department having jurisdiction over the lands be, and he hereby is, authorized and empowered, under general regulations to be fixed by him, to grant an easement for rights of way, for a period not exceeding fifty years from the date of the issuance of such grant, over, across, and upon the public lands, national forests, and reservations of the United States for electrical poles and lines for the transmission and distribution of electrical power, and for poles and lines for telephone and telegraph purposes, to the extent of twenty feet on each side of the center line of such electrical, telephone and telegraph lines and poles, to any citizen, association, or corporation of the United States, where it is intended by such to exercise the right of way herein granted for any one or more of the purposes herein named: *Provided*, That such right of way shall be allowed within or through any national park, national forest, military, Indian or any other reservation only upon the approval of the chief officer of the department under whose supervision or control such reservation falls, and upon a finding by him that the same is not incompatible with the public interest: *Provided*, That all or any part of such right of way may be forfeited and annulled by declaration of the head of the department having jurisdiction over the lands for nonuse for a period of two years or for abandonment.

That any citizen, association, or corporation of the United States to whom there has heretofore been issued a permit for any of the purposes specified herein under any existing law, may obtain the benefit of this Act upon the same terms and conditions as shall be required of citizens, associations, or corporations hereafter making application under the provisions of this statute.

1912—Act of March 11, 1912 (37 Stat. L., 74)—An Act To amend an Act entitled " An Act granting to certain employees of the United States the right to receive from it compensation for injuries sustained in the course of their employment," approved May thirtieth, nineteen hundred and eight.

[SECTION 1]. That the provisions of the Act approved May thirtieth, nineteen hundred and eight, entitled " An Act granting to certain employees of the United States the right to receive from it compensation for injuries sustained in the course of their employment," shall, in addition to the classes of persons therein designated, be held to apply to any artisan, laborer, or other employee engaged in any hazardous work under . . . the Forestry Service of the United States: *Provided,* That this Act shall not be held to embrace any case arising prior to its passage.[16]

1912—Act of August 10, 1912 (37 Stat. L., 269, 287, 294)—
An Act Making appropriations for the Department of Agriculture for the fiscal year ending June thirtieth, nineteen hundred and thirteen.

* * * *

That the Secretary of Agriculture, under such rules and regulations as he shall establish, is hereby authorized and directed to sell at actual cost, to homestead settlers and farmers, for their domestic use, the mature, dead, and down timber in national forests, but it is not the intent of this provision to restrict the authority of the Secretary of Agriculture to permit the free use of timber as provided in the Act of June fourth, eighteen hundred and ninety-seven.

* * * *

. . . That hereafter employees of the Division of Accounts and Disbursements may be detailed by the Secretary of Agriculture for accounting and disbursing work in any of the bureaus and offices of the department for duty in or out of the city of Washington, and employees of the bureaus and offices of the department may also be detailed to the Division of Accounts and Disbursements for duty in or out of the city of Washington, traveling expenses of employees so detailed to be paid from the appropriation of the bureau or office in connection with which such travel is performed.

1913—Act of March 4, 1913 (37 Stat. L., 828, 843)—An Act Making appropriations for the Department of Agriculture for the fiscal year ending June thirtieth, nineteen hundred and fourteen.

* * * *

. . . hereafter the Secretary of Agriculture, whenever he may deem it necessary for the protection of the national forests from fire, may permit the use of timber free of charge for the construction of telephone lines. . . .

---

[16] Superseded by act of September 7, 1916 (39 Stat. L., 742), "An Act to provide compensation for employees of the United States suffering injuries while in the performance of their duties, and for other purposes," as amended—in Section 20—by the act of June 13, 1922 (42 Stat. L., 650).

That hereafter an additional ten per centum of all moneys received from the national forests during each fiscal year shall be available at the end thereof, to be expended by the Secretary of Agriculture for the construction and maintenance of roads and trails within the national forests in the States from which such proceeds are derived; but the Secretary of Agriculture may, whenever practicable, in the construction and maintenance of such roads, secure the coöperation or aid of the proper State or Territorial authorities in the furtherance of any system of highways of which such roads may be made a part.[17]

* * * *

That hereafter the Secretary of Agriculture is authorized to reimburse owners of horses, vehicles, and other equipment lost, damaged, or destroyed while being used for necessary fire fighting, trail, or official business, such reimbursement to be made from any available funds in the appropriation to which the hire of such equipment is properly chargeable.[18]

* * * *

That hereafter the employees of the Forest Service who are assigned to permanent duty in Alaska may, in the discretion of the Secretary of Agriculture, without additional expense to the Government, be granted leave of absence not to exceed thirty days in any one year, which leave may, in exceptional and meritorious cases where such an employee is ill, be extended, in the discretion of the Secretary of Agriculture, not to exceed thirty days additional in any one year.

1914—Act of June 30, 1914 (38 Stat. L., 415, 430)—An Act Making appropriations for the Department of Agriculture for the fiscal year ending June thirtieth, nineteen hundred and fifteen.

* * * *

That hereafter all moneys received as contributions toward coöperative work in forest investigations, or the protection and improvement of the national forests, shall be covered into the Treasury and shall constitute a special fund, which is hereby appropriated and made available until expended, as the Secretary of Agriculture may direct, for the payment of the expenses of said investigations, protection, or improvements by the Forest Service, and for refunds to the contributors of amounts heretofore or hereafter paid in by them in excess of their share of the cost of said investigations, protection, or improvements: *Provided,* That annual report shall be made to Congress of all such moneys so received as contributions for such coöperative work.

[17] See similar clause in act of August 10, 1912 (37 Stat. L., 269, 288), covering 1913 only.
[18] An appropriation item for this purpose was previously contained in the deficiency appropriation act for 1911—act of March 4, 1911 (36 Stat. L., 1289, 1312).

1915—Act of March 4, 1915 (38 Stat. L., 1086, 1100)—An Act Making appropriations for the Department of Agriculture for the fiscal year ending June thirtieth, nineteen hundred and sixteen.

* * * *

That hereafter the Secretary of Agriculture, under regulations to be prescribed by him, is hereby authorized to permit the Navy Department to take from the national forests such earth, stone, and timber for the use of the Navy as may be compatible with the administration of the national forests for the purposes for which they are established, and also in the same manner to permit the taking of earth, stone, and timber from the national forests for the construction of Government railways and other Government works in Alaska: *Provided*, That the Secretary of Agriculture shall submit with his annual estimates a report of the quantity and market value of earth, stone, and timber furnished as herein provided.

1916—Act of July 11, 1916 (39 Stat. L., 355, 358)—An Act To provide that the United States shall aid the States in the construction of rural post roads, and for other purposes.[19]

SEC. 8. That there is hereby appropriated and made available until expended, out of any moneys in the National Treasury not otherwise appropriated, the sum of $1,000,000 for the fiscal year ending June thirtieth, nineteen hundred and seventeen, and each fiscal year thereafter, up to and including the fiscal year ending June thirtieth, nineteen hundred and twenty-six, in all $10,000,000, to be available until expended under the supervision of the Secretary of Agriculture, upon request from the proper officers of the State, Territory, or county for the survey, construction, and maintenance of roads and trails within or only partly within the national forests, when necessary for the use and development of resources upon which communities within and adjacent to the national forests are dependent: *Provided*, That the State, Territory, or county shall enter into a coöperative agreement with the Secretary of Agriculture for the survey, construction, and maintenance of such roads or trails upon a basis equitable to both the State, Territory, or county, and the United States: *And provided also*, That the aggregate expenditures in any State, Territory, or county shall not exceed ten per centum of the value, as determined by the Secretary of Agriculture, of the timber and forage resources which are or will be available for income upon the national forest lands within the respective county or counties wherein the roads or trails will be con-

[19] See act of November 9, 1921 (42 Stat. L., 212).

structed; and the Secretary of Agriculture shall make annual report to Congress of the amounts expended hereunder.

That immediately upon the execution of any coöperative agreement hereunder the Secretary of Agriculture shall notify the Secretary of the Treasury of the amount to be expended by the United States within or adjacent to any national forest thereunder, and beginning with the next fiscal year and each fiscal year thereafter the Secretary of the Treasury shall apply from any and all revenues from such forest ten per centum thereof to reimburse the United States for expenditures made under such agreement until the whole amount advanced under such agreement shall have been returned from the receipts from such national forest.

1916—Act of August 11, 1916 (39 Stat. L., 446, 462, 476)—
An Act Making appropriations for the Department of Agriculture for the fiscal year ending June thirtieth, nineteen hundred and seventeen, and for other purposes.

* * * *

That hereafter deposits may be received from timber purchasers in such sums as the Secretary of Agriculture may require to cover the cost to the United States of disposing of brush and other debris resulting from cutting operations in sales of national forest timber; such deposits shall be covered into the Treasury and shall constitute a special fund, which is hereby appropriated and made available until expended, as the Secretary of Agriculture may direct to pay the cost of such work and to make refunds to the depositors of amounts deposited by them in excess of such cost.[20]

* * * *

The Secretary of Agriculture is authorized, under general regulations to be prescribed by him, to permit the prospecting, development, and utilization of the mineral resources of the lands acquired under the Act of March first, nineteen hundred and eleven (Thirty-sixth Statutes, page nine hundred and sixty-one), known as the Weeks law, upon such terms and for specified periods or otherwise, as he may deem to be for the best interests of the United States; and all moneys received on account of charges, if any, made under this Act shall be disposed of as is provided, by existing law for the disposition of receipts from national forests.

* * * *

That the President of the United States is hereby authorized to designate such areas on any lands which have been, or which may hereafter be, purchased by the United States under the provisions of the

[20] See act of March 3, 1925 (43 Stat. L., 1132), Section 1.

Act of March first, nineteen hundred and eleven (Thirty-sixth Statutes at Large, page nine hundred and sixty-one) entitled "An Act to enable any State to coöperate with any other State or States, or with the United States, for the protection of watersheds of navigable streams, and to appoint a commission for the acquisition of lands for the purpose of conserving the navigability of navigable streams," and Acts supplementary thereto and amendatory thereof, as should, in his opinion, be set aside for the protection of game animals, birds, or fish; and whoever shall hunt, catch, trap, willfully disturb or kill any kind of game animal, game or nongame bird, or fish, or take the eggs of any such bird on any lands so set aside, or in or on the waters thereof, except under such general rules and regulations as the Secretary of Agriculture may from time to time prescribe, shall be fined not more than $500 or imprisoned not more than six months, or both.[21]

1917—Act of March 4, 1917 (39 Stat. L., 1134, 1149)—An Act Making appropriations for the Department of Agriculture for the fiscal year ending June thirtieth, nineteen hundred and eighteen.

* * * *

. . . *Provided*, That hereafter, all moneys received on account of permits for hunting, fishing, or camping, on lands acquired under authority of said Act, or any amendment or extension thereof, shall be disposed of as is provided by existing law for the disposition of receipts from national forests.

1919—Act of February 28, 1919 (40 Stat. L., 1189, 1201)— An Act Making appropriations for the service of the Post Office Department for the fiscal year ending June 30, 1920, and for other purposes.

* * * *

Sec. 7. That the Secretary of War be, and he is hereby, authorized in his discretion to transfer to the Secretary of Agriculture all available war material, equipment, and supplies not needed for the purposes of the War Department, but suitable for use in the improvement of highways, and that the same be distributed among the highway departments of the several States to be used on roads constructed in whole or in part by Federal aid, such distribution to be made upon a value basis of distribution the same as provided by the Federal aid road Act, approved July 11, 1916: *Provided*, That the Secretary of Agriculture, at his discretion, may reserve from such distribution not

[21] Under " Miscellaneous " section of the act. See also proviso under act of March 4, 1917 (39 Stat. L., 1134, 1150), and footnote under act of January 24, 1905 (33 Stat. L., 614).

to exceed 10 per centum of such material, equipment and supplies for use in the construction of national forest roads or other roads constructed under his direct supervision.

SEC. 8. That there is hereby appropriated, out of any money to the Treasury not otherwise appropriated, for the fiscal year ending June 30, 1919, the sum of $3,000,000 for the fiscal year ending June 30, 1920, the sum of $3,000,000, and for the fiscal year ending June 30, 1921, the sum of $3,000,000 available until expended by the Secretary of Agriculture in coöperation with the proper officials of the State, Territory, insular possession, or county, in the survey, construction, and maintenance of roads and trails within or partly within the national forests, when necessary for the use and development of resources of the same or desirable for the proper administration, protection, and improvement of any such forest. Out of the sums so appropriated the Secretary of Agriculture may, without the coöperation of such officials, survey, construct, and maintain any road or trail within a national forest which he finds necessary for the proper administration, protection, and improvement of such forest, or which in his opinion is of national importance. In the expenditure of this fund for labor preference shall be given, other conditions being equal, to honorably discharged soldiers, sailors, and marines.

The Secretary of Agriculture shall make annual report to Congress of the amounts expended hereunder.

1919—Act of July 24, 1919 (41 Stat. L., 234, 247, 270)—An Act Making appropriations for the Department of Agriculture for the fiscal year ending June 30, 1920.

* * * *

General Expenses, Forest Service: . . . to pay all expenses . . . including the payment of rewards under regulations of the Secretary of Agriculture for information leading to the arrest and conviction for violation of the laws and regulations relating to fires in or near national forests, or for the unlawful taking of, or injury to, Government property.[22]

* * * *

That hereafter in carrying on the activities of the Department of Agriculture involving coöperation with State, county and municipal agencies, associations of farmers, individual farmers, universities, colleges, boards of trade, chambers of commerce, or other local associations of business men, business organizations, and individuals within the State, Territory, district or insular possession in which such activities are to be carried on, moneys contributed from such outside sources, except in the case of the authorized activities of the Forest Service, shall be paid only through the Secretary of Agriculture or through State, county or municipal agencies, or local farm bureaus or like organizations, coöperating for the purpose with the Secretary of Agriculture.

[22] Repeated in all subsequent appropriation acts.

The officials and the employees of the Department of Agriculture engaged in the activities described in the preceding paragraph and paid in whole or in part out of funds contributed as provided therein, and the persons, corporations, or associations making contributions as therein provided, shall not be subject to the proviso contained in the Act making appropriations for the legislative, executive, and judicial expenses of the Government for the fiscal year ending June 30, 1918, and for other purposes, approved March 3, 1917, in Thirty-ninth Statutes at Large, at page 1106; nor shall any official or employee engaged in the coöperative activities of the Forest Service, or the persons, corporations, or associations contributing to such activities be subject to the said proviso.[23]

1920—Act of June 10, 1920 (41 Stat. L., 1063)—An Act To create a Federal Power Commission; to provide for the improvement of navigation; the development of water power; the use of the public lands in relation thereto, and to repeal Section 18 of the River and Harbor Appropriation Act, approved August 8, 1917, and for other purposes.[24]

[SECTION 1]. That a commission is hereby created . . . to be known as the Federal Power Commission . . . which shall be composed of the Secretary of War, the Secretary of the Interior, and the Secretary of Agriculture . . .

* * * *

SEC. 2. . . . The work of the commission shall be performed by and through the Departments of War, Interior, and Agriculture and their engineering, technical, clerical, and other personnel except as may be otherwise provided by law.

* * * *

SEC. 3. That the words defined in this section shall have the following meanings for the purposes of this Act, to wit:

* * * *

" Reservations " means . . . national forests . . .

* * * *

SEC. 4. That the commission is hereby authorized and empowered

* * * *

(d) To issue licenses to citizens of the United States, or to any association of such citizens, or to any corporation organized under the

[23] Under " Miscellaneous " section of the act. The proviso referred to prohibits government employees receiving salaries for their services in addition to their regular compensations from the government.

[24] The reorganization of the Federal Power Commission is provided for in the Act of June 23, 1930 (Public No. 412, 71st Congress).

laws of the United States or any State thereof, or to any State, or municipality for the purpose of constructing, operating, and maintaining dams, water conduits, reservoirs, power houses, transmission lines, or other project works necessary or convenient for the development and improvement of navigation, and for the development, transmission, and utilization of power across, along, from or in any of the navigable waters of the United States, or upon any part of the public lands and reservations of the United States (including the Territories), or for the purpose of utilizing the surplus water or water power from any Government dam, except as herein provided: *Provided,* That licenses shall be issued within any reservation only after a finding by the commission that the license will not interfere or be inconsistent with the purpose for which such reservation was created or acquired, and shall be subject to and contain such conditions as the Secretary of the department under whose supervision such reservation falls shall deem necessary for the adequate protection and utilization of such reservation.

* * * *

SEC. 6. That licenses under this act shall be issued for a period not exceeding fifty years.

* * * *

SEC. 10. That all licenses issued under this Act shall be on the following conditions:

* * * *

(e) That the licensee shall pay to the United States reasonable annual charges . . .

* * * *

SEC. 17. That all proceeds from any Indian reservation shall be placed to the credit of the Indians of such reservation. All other charges arising from licenses hereunder shall be paid into the Treasury of the United States, subject to the following distribution: Twelve and one-half per centum thereof is hereby appropriated to be paid into the Treasury of the United States and credited to " Miscellaneous receipts "; 50 per centum of the charges arising from licenses hereunder for the occupancy and use of public lands, national monuments, national forests, and national parks shall be paid into, reserved, and appropriated as a part of the reclamation fund created by the Act of Congress known as the Reclamation Act, approved June 17, 1902; and 37½ per centum of the charges arising from licenses hereunder for the occupancy and use of national forests, national parks, public lands, and national monuments, from development within the boundaries of any State shall be paid by the Secretary of the Treasury to such State; and 50 per centum of the charges arising from all other licenses hereunder is hereby reserved and appropriated as a special fund in the Treasury to be expended under the direction of the Secretary of War in the maintenance and operation of dams and other navigation structures owned by the United States or in the construction, maintenance, or operation of headwater or other improvements of navigable waters of the United States.

* * * *

SEC. 29. That all Acts or parts of Acts inconsistent with this Act are hereby repealed: *Provided,* That nothing herein contained shall be held or construed to modify or repeal any of the provisions of the Act of Congress approved December 19, 1913, granting certain rights of way to the city and county of San Francisco, in the State of California: *Provided further,* That section 18 of an Act making appropriations for the construction, repair, and preservation of certain public works on rivers and harbors, and for other purposes, approved August 8, 1917, is hereby repealed.

SEC. 30. That the short title of this Act shall be " The Federal Water Power Act."

1921—Act of November 9, 1921 (42 Stat. L., 212)—An Act To amend the Act entitled "An Act to provide that the United States shall aid the States in the construction of rural post roads, and for other purposes," approved July 11, 1916, as amended and supplemented, and for other purposes.

[SECTION 1]. That this Act may be cited as the Federal Highway Act.

SEC. 2. . . . The term " forest roads " means roads wholly or partly within or adjacent to and serving the national forests.

* * * *

SEC. 15. That within two years after this Act takes effect the Secretary of Agriculture shall prepare, publish, and distribute a map showing the highways and forest roads that have been selected and approved as a part of the primary or interstate, and the secondary or intercounty systems, and at least annually thereafter shall publish supplementary maps showing his program and the progress made in selection, construction, and reconstruction.

* * * *

SEC. 17. That if the Secretary of Agriculture determines that any part of the public lands or reservations of the United States is reasonably necessary for the right of way of any highway or forest road or as a source of materials for the construction or maintenance of any such highway or forest road adjacent to such lands or reservations, the Secretary of Agriculture shall file with the Secretary of the department supervising the administration of such lands or reservation a map showing the portion of such lands or reservations which it is desired to appropriate.

If within a period of four months after such filing the said Secretary shall not have certified to the Secretary of Agriculture that the proposed appropriation of such land or material is contrary to the public interest or inconsistent with the purposes for which such land or materials have been reserved, or shall have agreed to the appropriation and transfer under conditions which he deems necessary for the adequate protection and utilization of the reserve, then such land and materials may be appropriated and transferred to the State highway department for such purposes and subject to the conditions so specified.

If at any time the need for any such lands or materials for such purposes shall no longer exist, notice of the fact shall be given by the State highway department to the Secretary of Agriculture, and such lands or materials shall immediately revert to the control of the Secretary of the department from which they had been appropriated.

* * * *

SEC. 21. . . . That any sums apportioned to any State under the provisions of the Act entitled "An Act to provide that the United States shall aid the States in the construction of rural post roads, and for other purposes," approved July 11, 1916, and all Acts amendatory thereof and supplemental thereto, shall be available for expenditure in that State for the purpose set forth in such Acts until two years after the close of the respective fiscal years for which any such sums become available, and any amount so apportioned remaining unexpended at the end of the period during which it is available for expenditure under the terms of such Acts shall be reapportioned according to the provisions of the Act entitled "An Act to provide that the United States shall aid the States in the construction of rural post roads, and for other purposes," approved July 11, 1916: *And provided further,* That any amount apportioned under the provisions of this Act unexpended at the end of the period during which it is available for expenditures under the terms of this section shall be reapportioned within sixty days thereafter to all the States in the same manner and on the same basis, and certified to the Secretary of the Treasury and the State highway departments in the same way as if it were being apportioned under this Act for the first time.

* * * *

SEC. 23. That out of the moneys in the Treasury not otherwise appropriated, there is hereby appropriated for the survey, construction, reconstruction, and maintenance of forest roads and trails, the sum of $5,000,000 for the fiscal year ending June 30, 1922, available immediately and until expended, and $10,000,000 for the fiscal year ending June 30, 1923, available until expended.

(a) Fifty per centum, but not to exceed $3,000,000 for any one fiscal year, of the appropriation made or that may hereafter be made for expenditure under the provisions of this section shall be expended under the direct supervision of the Secretary of Agriculture in the survey, construction, reconstruction, and maintenance of roads and trails of primary importance for the protection, administration, and utilization of the national forests, or when necessary, for the use and development of resources upon which communities within or adjacent to the national forests are dependent, and shall be apportioned among the several States, Alaska, and Porto Rico by the Secretary of Agriculture, according to the relative needs of the various national forests, taking into consideration the existing transportation facilities, value of timber, or other resources served, relative fire danger, and comparative difficulties of road and trail construction.

The balance of such appropriations shall be expended by the Secretary of Agriculture in the survey, construction, reconstruction, and maintenance of forest roads of primary importance to the State, counties, or communities within, adjoining, or adjacent to the national forests, and shall be prorated and apportioned by the Secretary of Agriculture for expenditures in the several States, Alaska, and Porto Rico, according to the area and value of the land owned by the Government within the national forests therein as determined by the Secretary of Agriculture from such information, investigation, sources, and departments as the Secretary of Agriculture may deem most accurate.

(b) Coöperation of Territories, States, and civil subdivisions thereof may be accepted but shall not be required by the Secretary of Agriculture.

(c) The Secretary of Agriculture may enter into contracts with any Territory, State, or civil subdivision thereof for the construction, reconstruction, or maintenance of any forest road or trail or part thereof.

(d) Construction work on forest roads or trails estimated to cost $5,000 or more per mile, exclusive of bridges, shall be advertised and let to contract.

If such estimated cost is less than $5,000 per mile, or if, after proper advertising, no acceptable bid is received, or the bids are deemed excessive, the work may be done by the Secretary of Agriculture on his own account; and for such purpose the Secretary of Agriculture may purchase, lease, hire, rent, or otherwise obtain all necessary supplies, materials, tools, equipment, and facilities required to perform the work.

The appropriation made in this section or that may hereafter be made for expenditure under the provisions of this section may be expended for the purpose herein authorized and for the payment of wages, salaries, and other expenses for help employed in connection with such work.

1922—Act of March 20, 1922 (42 Stat. L., 465)—An Act To consolidate national forest lands.

[SECTION 1]. That when the public interests will be benefited thereby, the Secretary of the Interior be, and hereby is, authorized in his discretion to accept on behalf of the United States title to any lands within the exterior boundaries of the national forests which, in the opinion of the Secretary of Agriculture, are chiefly valuable for national forest purposes, and in exchange therefor may patent not to exceed an equal value of such national forest land, in the same State, surveyed and nonmineral in character, or the Secretary of Agriculture may authorize the grantor to cut and remove an equal value of timber within the national forests of the same States; the values in each case to be determined by the Secretary of Agriculture: *Provided,* That before any such exchange is effected notice of the contemplated exchange reciting the lands involved shall be published once each week for four successive weeks in some newspaper of general circulation in the county or counties in which may be situated the lands to be accepted, and in some like newspaper published in any county in which may be situated any lands or timber to be given in such exchange. Timber given in such exchanges shall be cut and removed under the laws and regulations relating to the national forests, and under the direction and supervision and in accordance with the requirements of the Secretary of Agriculture. Lands conveyed to the United States under this Act shall, upon acceptance of title, become parts of the national forest within whose exterior boundaries they are located.

[Added by an act of February 28, 1925 (43 Stat. L., 1090), An Act To amend an Act entitled "An Act to consolidate national forest lands."]

SEC. 2. Either party to an exchange may make reservations of timber, minerals, or easements, the values of which shall be duly considered in determining the values of the exchanged lands. Where reservations are made in lands conveyed to the United States the right to enjoy them shall be subject to such reasonable conditions respecting ingress and egress and the use of the surface of the land as may be deemed necessary by the Secretary of Agriculture; where mineral reservations are made in lands conveyed by the United States it shall be so stipulated in the patents, and that any person who acquires the right to mine and remove the reserved deposits may enter and occupy so much of the surface as may be required for all purposes incident to the mining and removal of the minerals therefrom, and may mine and remove such minerals upon payment to the owner of the surface for damages caused to the land and improvements thereon: *Provided.* That all property, rights, easements, and benefits authorized by this section to be retained by or reserved to owners of lands conveyed to the United States shall be subject to the tax laws of the States where such lands are located.

1922—Act of May 11, 1922 (42 Stat. L., 507, 521)—An Act Making appropriations for the Department of Agriculture for the fiscal year ending June 30, 1923, and for other purposes.

* * * *

. . . *Provided further,* That hereafter [25] no part of any funds appropriated for the Forest Service shall be used to pay the transportation

[25] These provisos first appeared in the act of May 23, 1908 (35 Stat. L., 259), "An Act making appropriations for the Department of Agriculture for the fiscal year ending June thirtieth, nineteen hundred and nine," and were repeated in each subsequent appropriation act until made general and permanent by the language here used.

or traveling expenses of any forest officer or agent except he be traveling on business directly connected with the Forest Service and in furtherance of the works, aims, and objects specified and authorized by law: *And provided further,* That hereafter [25] no part of any funds appropriated for the Forest Service shall be paid or used for the purpose of paying for, in whole or in part, the preparation or publication of any newspaper or magazine article, but this shall not prevent the giving out to all persons, without discrimination, including newspapers and magazine writers and publishers, of any facts or official information of value to the public.

1924—Act of June 7, 1924 (43 Stat. L., 653)—An Act To provide for the protection of forest lands, for the reforestation of denuded areas, for the extension of national forests, and for other purposes, in order to promote the continuous production of timber on lands chiefly suitable therefor.[26]

[SECTION 1]. That the Secretary of Agriculture is hereby authorized and directed, in coöperation with appropriate officials of the various States or other suitable agencies, to recommend for each forest region of the United States such systems of forest fire prevention and suppression as will adequately protect the timbered and cut-over lands therein with a view to the protection of forest and water resources and the continuous production of timber on lands chiefly suitable therefor.

SEC. 2. That if the Secretary of Agriculture shall find that the system and practice of forest fire prevention and suppression provided by any State substantially promotes the objects described in the foregoing section, he is hereby authorized and directed, under such conditions as he may determine to be fair and equitable in each State, to coöperate with appropriate officials of each State, and through them with private and other agencies therein, in the protection of timbered and forest-producing lands from fire. In no case other than for preliminary investigations shall the amount expended by the Federal Government in any State during any fiscal year, under this section, exceed the amount expended by the State for the same purpose during the same fiscal year, including the expenditures of forest owners or operators which are required by State law or which are made in pursuance of the forest protection system of the State under State supervision and for which in all cases the State renders satisfactory accounting. In the coöperation extended to the several States due consideration shall be given to the protection of watersheds of navigable streams, but such coöperation may, in the discretion of the Secretary of Agriculture, be extended to any timbered or forest producing lands *or watersheds from which water is secured for domestic use or irrigation*[27] within the coöperating States.

[26] The Clarke-McNary Act.
[27] Words in italics added by act of March 3, 1925 (43 Stat. L., 1127).

Sec. 3. That the Secretary of Agriculture shall expend such portions of the appropriations authorized herein as he deems advisable to study the effects of tax laws, methods, and practices upon forest perpetuation, to coöperate with appropriate officials of the various States or other suitable agencies in such investigations and in devising tax laws designed to encourage the conservation and growing of timber, and to investigate and promote practical methods of insuring standing timber on growing forests from losses by fire and other causes. There is hereby authorized to be appropriated annually, out of any money in the Treasury not otherwise appropriated, not more than $2,500,000, to enable the Secretary of Agriculture to carry out the provisions of sections 1, 2, and 3 of this Act.

Sec. 4. That the Secretary of Agriculture is hereby authorized and directed to coöperate with the various States in the procurement, production and distribution of forest-tree seeds and plants, for the purpose of establishing wind breaks, shelter belts, and farm wood lots upon denuded or nonforested lands within such coöperating States, under such conditions and requirements as he may prescribe to the end that forest-tree seeds or plants so procured, produced, or distributed shall be used effectively for planting denuded or nonforested lands in the coöperating States and growing timber thereon: *Provided,* That the amount expended by the Federal Government in coöperation with any State during any fiscal year for such purposes shall not exceed the amount expended by the State for the same purposes during the same fiscal year. There is hereby authorized to be appropriated annually, out of any money in the Treasury not otherwise appropriated, not more than $100,000, to enable the Secretary of Agriculture to carry out the provisions of this section.

Sec. 5. That the Secretary of Agriculture is hereby authorized and directed, in coöperation with appropriate officials of the various States or, in his discretion, with other suitable agencies, to assist the owners of farms in establishing, improving, and renewing woodlots, shelter belts, windbreaks, and other valuable forest growth, and in growing and renewing useful timber crops: *Provided,* That, except for preliminary investigations, the amount expended by the Federal Government under this section in coöperation with any State or other coöperating agency during any fiscal year shall not exceed the amount expended by the State or other coöperating agency for the same purpose during the same fiscal year. There is hereby authorized to be appropriated annually out of any money in the Treasury not otherwise appropriated, not more than $100,000 to enable the Secretary of Agriculture to carry out the provisions of this section.

Sec. 6. That section 6 of the Act of March 1, 1911 (Thirty-sixth Statutes at Large, page 961), is hereby amended to authorize and direct the Secretary of Agriculture to examine, locate, and recommend for purchase such forested, cut-over or denuded lands within the watersheds of navigable streams as in his judgment may be necessary to the regulation of the flow of navigable streams or for the production of timber and to report to the National Forest Reservation Commission the results of such examination; but before any lands are purchased by the commission said lands shall be examined by the Secretary of Agriculture, in coöperation with the Director of the Geological Survey, and a report made by them to the commission showing that the control of such lands by the Federal Government will promote or protect the navigation of streams or by the Secretary of Agriculture showing that such control will promote the production of timber thereon.[28]

SEC. 7. That to enable owners of lands chiefly valuable for the growing of timber crops to donate or devise such lands to the United States in order to assure future timber supplies for the agricultural and other industries of the State or for other national forest purposes, the Secretary of Agriculture is hereby authorized, in his discretion, to accept [29] on behalf of the United States title to any such land so donated or devised, subject to such reservations by the donor of the present stand of merchantable timber or of mineral or other rights for a period not exceeding twenty years as the Secretary of Agriculture may find to be reasonable and not detrimental to the purposes of this section, and to pay out of any moneys appropriated for the general expenses of the Forest Service the cost of recording deeds or other expenses incident to the examination and acceptance of title. Any lands to which title is so accepted shall be in units of such size or so located as to be capable of economical administration as national forests either separately or jointly with other lands acquired under this section, or jointly with an existing national forest. All lands to which title is accepted under this section, shall, upon acceptance of title, become national forest lands, subject to all laws applicable to lands acquired under the Act of March 1, 1911 (Thirty-sixth Statutes at Large, page 961), and amendments thereto. In the sale of timber from national forest lands acquired under this section preference shall be given to applicants who will furnish the products desired therefrom to meet the necessities of citizens of the United States engaged in agriculture in the States in which such national forest is situated: *Provided,* That all property, rights, easements, and benefits authorized by this section to be retained by or reserved to owners of lands donated or devised to the United States shall be subject to the tax laws of the States where such lands are located.

SEC. 8. That the Secretary of Agriculture is hereby authorized to ascertain and determine the location of public lands chiefly valuable for stream-flow protection or for timber production, which can be economically administered as parts of national forests, and to report his findings to the National Forest Reservation Commission established under the Act of March 1, 1911 (Thirty-sixth Statutes at Large, page 961), and if the commission shall determine that the administration of

[28] See Section 6 of the Weeks Act.
[29] See the final clause of Section 5 of act of March 3, 1925 (43 Stat. L., 1132).

said lands by the Federal Government will protect the flow of streams used for navigation or for irrigation, or will promote a future timber supply, the President shall lay the findings of the Commission before the Congress of the United States.

SEC. 9. That the President, in his discretion, is hereby authorized to establish as national forests, or parts thereof, any lands within the boundaries of Government reservations, other than national parks, reservations for phosphate and other mineral deposits or water-power purposes, national monuments, and Indian reservations, which in the opinion of the Secretary of the department now administering the area and the Secretary of Agriculture are suitable for the production of timber, to be administered by the Secretary of Agriculture under such rules and regulations and in accordance with such general plans as may be jointly approved by the Secretary of Agriculture and the Secretary formerly administering the area, for the use and occupation of such lands and for the sale of products therefrom. That where such national forest is established on land previously reserved for the Army or Navy for purposes of national defense the land shall remain subject to the unhampered use of the War or Navy Department for said purposes, and nothing in this section shall be construed to relinquish the authority over such lands for purposes of national defense now vested in the Department for which the lands were formerly reserved. Any moneys available for the maintenance, improvement, protection, construction of highways and general administration of the national forests shall be available for expenditure on the national forests created under this section. All receipts from the sale of products from or for the use of lands in such national forests shall be covered into the Treasury as miscellaneous receipts, forest reserve fund, and shall be disposed of in like manner as the receipts from other national forests as provided by existing law. Any person who shall violate any rule or regulation promulgated under this section shall be guilty of a misdemeanor, and upon conviction thereof shall be fined not more than $500 or imprisoned for not more than one year, or both.

1925—Act of February 28, 1925 (43 Stat. L., 1090)—An Act To amend an Act entitled "An Act to consolidate national forest lands."

[SECTION 1]. The Act of March 20, 1922 (Forty-second Statutes at Large, page 465), entitled "An Act to consolidate national forest lands," be, and the same is hereby, amended by adding the following section thereto:

SEC. 2. Either party to an exchange may make reservations of timber, minerals, or easements, the values of which shall be duly considered in determining the values of the exchanged lands. Where reservations are made in lands conveyed to the United States the right to enjoy them shall be subject to such reasonable conditions respecting ingress and egress and the use of the surface of the land as may be deemed necessary by the Secretary of Agriculture; where mineral reservations are made in lands conveyed by the United States it shall be so stipulated in the patents, and that any person who acquires the right to mine and remove the reserved deposits may enter and occupy so much of the surface as may be required for all purposes incident to the mining and removal of the minerals therefrom, and may mine and remove such minerals upon payment to the owner of the surface for damages caused to the land and improvements thereon: *Provided,* That all property, rights, easements, and benefits authorized by this section to be retained by or reserved to owners of lands conveyed to the United States shall be subject to the tax laws of the States where such lands are located.

1925—Act of March 3, 1925 (43 Stat. L., 1132)—An Act To facilitate and simplify the work of the Forest Service, United States Department of Agriculture, and to promote reforestation.

[SECTION 1]. That all moneys received as contributions toward reforestation or for the administration or protection of lands within or near the national forests shall be covered into the Treasury and shall constitute a special fund, which is hereby authorized to be appropriated for the payment of the expenses of said reforestation, administration, or protection by the Forest Service, and for refunds to the contributors of amounts heretofore or hereafter paid in by or for them in excess of their share of the cost, but the United States shall not be liable for any damage incident to coöperation hereunder.

SEC. 2. That, in addition to buildings costing not to exceed $1,500 each, the Secretary of Agriculture, out of any moneys appropriated for the improvement or protection of the national forests, may construct, improve, or purchase during each fiscal year three buildings for national forest purposes at not to exceed $2,500 each, and three at not to exceed $2,000 each: *Provided,* That the cost of a water supply or sanitary system shall not be charged as a part of the cost of any building except those costing in excess of $2,000 each, and no such water supply and sanitary system shall cost in excess of $500.

SEC. 3. That the Act of June 6, 1900 (Thirty-first Statutes, page 661), is hereby amended to enable the Secretary of Agriculture, in his discretion, to sell, without advertisement, in quantities to suit applicants, at a fair appraisement, timber, cordwood, and other forest products not exceeding $500 in appraised value.

SEC. 4. That the Secretary of Agriculture is hereby authorized to furnish subsistence to employees of the Forest Service, to purchase personal equipment and supplies for them, and to make deductions therefor from moneys appropriated for salary payments or otherwise due such employees.

SEC. 5. That where no suitable Government land is available for national forest headquarters or ranger stations, the Secretary of Agriculture is hereby authorized to purchase such lands out of any funds appropriated for building improvements on the national forests, but not more than $2,500 shall be so expended in any one year; and to accept donations of land for any national forest purpose.

SEC. 6. That the Secretary of Agriculture is hereby authorized, in his discretion, to provide out of moneys appropriated for the general expenses of the Forest Service medicinal attention for employees of the Forest Service located at isolated situations, including the moving of such employees to hospitals or other places where medical assistance is available, and in case of death to remove the bodies of deceased employees, to the nearest place where they can be prepared for shipment or for burial.

1926—Act of April 13, 1926 (44 Stat. L., 242)—An Act To amend section 2 of the Act of June 7, 1924 (Forty-third Statutes at Large, page 653), as amended by the Act of March 3, 1925 (Forty-third Statutes at Large, page 1127), entitled "An Act to provide for the protection of forest lands, for the reforestation of denuded areas, for the extension of national forests, and for other purposes, in order to promote the continuous production of timber on lands chiefly suitable therefor."

[SECTION 1]. The second sentence of section 2 of the Act entitled "An Act to provide for the protection of forest lands, for the reforestation of denuded areas, for the extension of national forests, and for other purposes, in order to promote the continuous production of timber on lands chiefly suitable therefor," approved June 7, 1924 (Forty-third Statutes at Large, page 653), as amended by the Act of March 3, 1925 (Forty-third Statutes at Large, page 1127), is further amended by striking out the words " and for which in all cases the State renders satisfactory accounting " and substituting the following: " and the Secretary of Agriculture is authorized to make expenditures on the certificate of the State forester, the State director of extension, or similar State official having charge of the coöperative work for the State that State and private expenditures as provided for in this Act have been made," so that section 2 as amended will read as follows:

SEC. 2. If the Secretary of Agriculture shall find that the system and practice of forest-fire prevention and suppression provided by any State substantially promotes the objects described in the foregoing section he is hereby authorized and directed, under such conditions as he may determine to be fair and equitable in each State, to coöperate with appropriate officials of each State, and through them with private and other agencies, therein, in the protection of timbered and forest-producing lands from fire. In no other case other than for preliminary investigation shall the amount expended by the Federal Government in any State during any fiscal year, under this section, exceed the amount expended by the State for the same purpose during the same fiscal year, including the expenditures of forest owners or operators which are required by State law or which are made in pursuance of the forest-protection system of the State under State supervision, and the Secretary of Agriculture is authorized to make expenditures on the certificate of the State forester, the State director of extension, or similar State official having charge of the coöperative work for the State that State and private expenditures as provided for in this Act have been made. In the coöperation extended to the several States due consideration shall be given to the protection of watersheds of navigable streams, but such coöperation may, in the discretion of the Secretary of Agriculture, be extended to any timbered or forest-producing lands or watersheds from which water is secured for domestic use or irrigation within the coöperative States.

1926—Act of April 30, 1926 (44 Stat. L., 374)—An Act Amending the Act entitled "An Act providing for a comprehensive development of the park and playground system of the National Capital" approved June 6, 1924.

SECTION 1. (a) That . . . to preserve forests and natural scenery in and about Washington [and other purposes] . . . there is hereby constituted . . . the National Capital Park and Planning Commission composed of [among others] . . . the Chief of the Forest Service . . ."

1928—Act of April 30, 1928 (45 Stat. L., 468)—An Act Authorizing an appropriation to be expended under the provisions of section 7 of the Act of March 1, 1911, entitled "An Act to enable any State to coöperate with any other State or States, or with the United States, for the protection of the watersheds of navigable streams, and to appoint a commission for the acquisition of lands for the purpose of conserving the navigability of navigable rivers," as amended.[30]

That there is hereby authorized to be appropriated, out of any money in the United States Treasury not otherwise appropriated, to

[30] The Woodruff-McNary Act.

be expended under the provisions of section 7 of the Act of March 1, 1911 (Thirty-sixth Statutes, page 961), as amended by the Acts of March 4, 1913 (Thirty-seventh Statutes, page 828), June 30, 1914 (Thirty-eighth Statutes, page 441), and the Act of June 7, 1924 (Public, 270), available July 1, 1928, $2,000,000; available July 1, 1929, $3,000,000; available July 1, 1930, $3,000,000; in all for this period, $8,000,000, to be available until expended: *Provided*, That, except for the protection of the headwaters of navigable streams or the control and reduction of floods therein, no lands shall be purchased under the appropriations herein authorized in excess of one million acres in any one State.

1928—Act of May 22, 1928 (45 Stat. L., 699)—An Act To insure adequate supplies of timber and other forest products for the people of the United States, to promote the full use for timber growing and other purposes of forest lands in the United States, including farm wood lots and those abandoned areas not suitable for agricultural production, and to secure the correlation and the most economical conduct of forest research in the Department of Agriculture, through research in reforestation, timber growing, protection, utilization, forest economics, and related subjects, and for other purposes.[31]

[Section 1]. That the Secretary of Agriculture is hereby authorized and directed to conduct such investigations, experiments, and tests as he may deem necessary under sections 2 to 10, inclusive, in order to determine, demonstrate, and promulgate the best methods of reforestation and of growing, managing, and utilizing timber, forage, and other forest products, of maintaining favorable conditions of water flow and the prevention of erosion, of protecting timber and other forest growth from fire, insects, disease, or other harmful agencies, of obtaining the fullest and most effective use of forest lands, and to determine and promulgate the economic considerations which should underlie the establishment of sound policies for the management of forest land and the utilization of forest products: *Provided*, That in carrying out the provisions of this Act the Secretary of Agriculture may coöperate with individuals and public and private agencies, organizations, and institutions, and, in connection with the collection, investigation, and tests of foreign woods, he may also coöperate with individuals and public and private agencies, organizations, and institutions in other countries; and receive money contributions from

[31] The McSweeney-McNary Act.

coöperators under such conditions as he may impose, such contributions to be covered into the Treasury as a special fund which is hereby appropriated and made available until expended as the Secretary of Agriculture may direct, for use in conducting the activities authorized by this Act, and in making refunds to contributors: *Provided further,* That the cost of any building purchased, erected, or as improved in carrying out the purposes of this Act shall not exceed $2,500, exclusive in each instance of the cost of constructing a water supply or sanitary system and of connecting the same with any such building: *Provided further,* That the amounts specified in sections 2, 3, 4, 5, 6, 7, 8, and 10 of this Act are authorized to be appropriated up to and including the fiscal year 1938, and such annual appropriations as may thereafter be necessary to carry out the provisions of said sections are hereby authorized: *Provided further,* That during any fiscal year the amounts specified in sections 3, 4, and 5 of this Act making provision for investigations of forest tree and wood diseases, forest insects, and forest wild life, respectively, may be exceeded to provide adequate funds for special research required to meet any serious public emergency relating to epidemics: *And provided further,* That the provisions of this Act shall be construed as supplementing all other Acts relating to the Department of Agriculture, and except as specifically provided shall not limit or repeal any existing legislation or authority.

SEC. 2. That for conducting fire, silvicultural, and other forest investigations and experiments the Secretary of Agriculture is hereby authorized, in his discretion, to maintain the following forest experiment stations for the regions indicated, and in addition to establish and maintain one such station for the Intermountain region in Utah and adjoining States, one in Alaska, and one in the tropical possessions of the United States in the West Indies:

Northeastern forest experiment station, in New England, New York, and adjacent States;

Allegheny forest experiment station, in Pennsylvania, New Jersey, Delaware, Maryland, and in neighboring States;

Appalachian forest experiment station, in the southern Appalachian Mountains and adjacent forest regions;

Southern forest experiment station, in the Southern States;

Central States forest experiment station in Ohio, Indiana, Illinois, Kentucky, Missouri, Iowa, and in adjacent States;

Lake States forest experiment station, in the Lake States and adjoining States;

California forest experiment station, in California and in adjoining States;

Northern Rocky Mountain forest experiment station, in Idaho, Montana and adjoining States;

Northwestern forest experiment station, in Washington, Oregon, and adjoining States, and in Alaska;

Rocky Mountain forest experiment station, in Colorado, Wyoming, Nebraska, South Dakota, and in adjacent States; and

Southwestern forest experiment station, in Arizona, and New Mexico, and in adjacent States, and in addition to establish and maintain one such station for the intermountain region of Utah and adjoining States, one for Alaska, one in Hawaii, and one in the tropical possessions of the United States in the West Indies, and one additional station in the Southern States.

There is hereby authorized to be appropriated annually out of any money in the Treasury not otherwise appropriated, not more than $1,000,000 to carry out the provisions of this section.

SEC. 3. That for investigations of the diseases of forest trees and of diseases causing decay and deterioration of wood and other forest products, and for developing methods for their prevention and control at forest experiment stations, the Forest Products Laboratory, or elsewhere, there is hereby authorized to be appropriated annually, out of any money in the Treasury not otherwise appropriated, not more than $250,000.

SEC. 4. That for investigations of forest insects, including gypsy and browntail moths, injurious or beneficial to forest trees or to wood or other forest products, and for developing methods for preventing and controlling infestations, at forest experiment stations, the Forest Products Laboratory, or elsewhere, there is hereby authorized to be appropriated annually, out of any money in the Treasury not otherwise appropriated, not more than $350,000.

SEC. 5. That for such experiments and investigations as may be necessary in determining the life histories and habits of forest animals, birds, and wild life, whether injurious to forest growth or of value as supplemental resource, and in developing the best and most effective methods for their management and control at forest experiment stations, or elsewhere, there is hereby authorized to be appropriated annually, out of any money in the Treasury not otherwise appropriated, not more than $150,000.

SEC. 6. That for such investigations at forest experiment stations, or elsewhere, of the relationship of weather conditions to forest fires as may be necessary to make weather forecasts, there is hereby authorized to be appropriated annually, out of any money in the Treasury not otherwise appropriated, not more than $50,000.

SEC. 7. That for such experiments and investigations as may be necessary to develop improved methods of management, consistent with the growing of timber and the protection of watersheds, of forest ranges and of other ranges adjacent to the national forests, at forest or range experiment stations, or elsewhere, there is hereby authorized to be appropriated annually, out of any money in the Treasury not otherwise appropriated, not more than $275,000.

SEC. 8. That for experiments, investigations, and tests with respect to the physical and chemical properties and the utilization and preservation of wood and other forest products, including tests of wood and other fibrous material for pulp and paper making and such other experiments, investigations, and tests as may be desirable, at the Forest Products Laboratory or elsewhere, there is hereby authorized to be appropriated annually, out of any money in the Treasury not otherwise appropriated, not more than $1,000,000, and an additional appropriation of not more than $50,000 annually for similar experiments, investigations and tests of foreign woods and forest products important to the industries of the United States, including necessary field work in connection therewith.

SEC. 9. That the Secretary of Agriculture is hereby authorized and directed, under such plans as he may determine to be fair and equitable, to coöperate with appropriate officials of each State of the United States, and either through them or directly with private and other agencies, in making a comprehensive survey of the present and prospective requirements for timber and other forest products in the United States, and of timber supplies, including a determination of the present and potential productivity of forest land therein, and of such other facts as may be necessary in the determination of ways and means to balance the timber budget of the United States. There is hereby authorized to be appropriated annually, out of any money in the Treasury not otherwise appropriated, not more than $250,000: *Provided,* That the total appropriation of Federal funds under this section shall not exceed $3,000,000.

SEC. 10. That for such investigations of costs and returns and the possibility of profitable reforestation under different conditions in the different forest regions, of the proper function of timber growing in diversified agriculture and in insuring the profitable use of marginal land, in mining, transportation, and in other industries, of the most effective distribution of forest products in the interest of both consumer and timber grower, and for such other economic investigations of forest lands and forest products as may be necessary, there is hereby authorized to be appropriated annually, out of any money in the Treasury not otherwise appropriated, not more than $250,000.

1930—Act of May 27, 1930 (Public No. 272, Seventy-first Congress)—An Act making appropriations for the Department of Agriculture for the fiscal year ending June 30, 1931, and for other purposes.

* * * *

FOREST SERVICE

SALARIES AND GENERAL EXPENSES

To enable the Secretary of Agriculture to experiment and to make and continue investigations and report on forestry, national forests, forest fires, and lumbering, but no part of this appropriation shall be used for any experiment or test made outside the jurisdiction of the United States; to advise the owners of woodlands as to the proper care of the same; to investigate and test American timber and timber trees and their uses, and methods for the preservative treatment of timber; to seek, through investigations and the planting of native and foreign species, suitable trees for the treeless regions; to erect necessary buildings: *Provided,* That the cost of any building purchased, erected, or as improved, exclusive of the cost of constructing a water supply or sanitary system and of connecting the same with any such building, and exclusive of the cost of any tower upon which a lookout house may be erected, shall not exceed $2,500; to pay all expenses necessary to protect, administer, and improve the national forests, including tree planting in the forest reserves to prevent erosion, drift, surface wash, and soil waste and the formation of floods, and including the payment of rewards under regulations of the Secretary of Agriculture for information leading to the arrest and conviction for violation of the laws and regulations relating to fires in or near national forests, or for the unlawful taking of, or injury to, Government property; to ascertain the natural conditions upon and utilize the national forests; to transport and care for fish and game supplied to stock the national forests or the waters therein; to employ agents, clerks, assistants, and other labor required in practical forestry and in the administration of national forests in the city of Washington and elsewhere; to collate, digest, report and illustrate the results of experiments and investigations made by the Forest Service; to purchase necessary supplies, apparatus, office fixtures, law books, reference and technical books and technical journals for officers of the Forest Service stationed outside of Washington, and for medical supplies and services and other assistance necessary for the immediate relief of artisans, laborers, and other employees engaged in any hazardous work under the Forest Service; to pay freight, express, telephone, and telegraph charges; for electric light and power, fuel, gas, ice, and washing towels, and official traveling and other necessary expenses, including traveling expenses for legal and fiscal officers while performing Forest Service work; and for rent outside of the District of Columbia, as follows:

For necessary expenses for general administrative purposes, including the salary of the Chief Forester and other personal services in the District of Columbia, $362,230.

NATIONAL FOREST ADMINISTRATION

For the employment of forest supervisors, deputy forest supervisors, forest rangers, forest guards, and administrative clerical assistants on the national forests, and for additional salaries and field-station expenses, including the maintenance of nurseries, collecting seed, and planting, necessary for the use, maintenance, improvement, and protection of the national forests, and of additional national forests created or to be created under section 11 of the Act of March 1, 1911 (U. S. C., title 16, sec. 521), and under the Act of June 7, 1924 (U. S. C., title 16, secs. 471, 499, 505, 564-570), and lands under contract for purchase or for the acquisition of which condemnation proceedings have been instituted for the purposes of said acts, and for necessary miscellaneous expenses incident to the general administration of the Forest Service and of the national forests:

In national forest district 1, Montana, Washington, Idaho, and South Dakota, $1,449,026: *Provided,* That the Secretary of Agriculture is authorized to use not to exceed $200 in caring for the graves of fire fighters buried at Wallace, Idaho; Priest River, Idaho; Newport, Washington; and Saint Maries, Idaho;

In national forest district 2, Colorado, Wyoming, South Dakota, Nebraska, and Oklahoma, $706,872: *Provided,* That not to exceed $1000 of this appropriation may be expended for the maintenance of the herd of long-horned cattle on the Wichita National Forest;

In national forest district 3, Arizona and New Mexico, $692,594;

In national forest district 4, Utah, Idaho, Wyoming, Nevada, Arizona, and Colorado, $910,514;

In national forest district 5, California and Nevada, $1,211,807;

In national forest district 6, Washington, Oregon, and California, $1,223,448;

In national forest district 7, Arkansas, Alabama, Florida, Georgia, South Carolina, North Carolina, Pennsylvania, Tennessee, Virginia, West Virginia, New Hampshire, Maine, Porto Rico, Maryland, New York, New Jersey, Kentucky, Louisiana, Mississippi, Vermont, and Illinois, $525,154;

In national forest district 8, Alaska, $139,007: *Provided,* That of the sum herein appropriated, $16,000 shall be available only for the purchase or construction of a boat for use in Alaska;

In national forest district 9, Michigan, Minnesota, and Wisconsin, $126,578;

In all, for the use, maintenance, improvement, protection, and general administration of the national forests, $6,985,000: *Provided,* That the foregoing amounts appropriated for such purposes shall be available interchangeably in the discretion of the Secretary of Agriculture for the necessary expeditures for the fire protection and other unforeseen exigencies: *Provided further,* That the amount so interchanged shall not exceed in the aggregate 10 per centum of all the amounts so appropriated.

For fighting and preventing forest fires on or threatening the national forests and for the establishment and maintenance of a patrol to prevent trespass and to guard against and check fires upon the lands revested in the United States by the Act approved June 9, 1916 (39 Stat., p.218), and the lands known as the Coos Bay Wagon Road lands involved in the case of Southern Oregon Company against United States (numbered 2711), in the Circuit Court of Appeals of the Ninth Circuit, $100,000, which amount shall be immediately available;

For coöperation with the War Department, or for contract airplane service, in the maintenance and operation of an airplane patrol to prevent and suppress forest fires on national forests and adjacent lands, $50,000; *Provided*, That no part of this appropriation shall be used for the purchase of land or airplanes.

For the selection, classification, and segregation of lands within the boundaries of national forests that may be opened to homestead settlement and entry under the homestead laws applicable to the national forests; for the examination and appraisal of lands in effecting exchanges authorized by law and for the survey thereof by metes and bounds or otherwise, by employees of the Forest Service, under the direction of the Commissioner of the General Land Office; and for the survey and platting of certain lands, chiefly valuable for agriculture, now listed or to be listed within the national forests, under the Act of June 11, 1906 (U. S. C., title 16, secs. 506-509), the Act of August 10, 1912 (U. S. C., title 16, sec. 506), and the Act of March 3, 1899 (U. S. C., title 16, sec. 488), as provided by the Act of March 4, 1913 (U. S. C., title 16, sec. 512), $52,500;

For the construction of sanitary facilities and for fire-preventive measures on public camp grounds within the national forests when necessary for the protection of the public health or the prevention of forest fires, $57,000;

For the purchase and maintenance of necessary field, office, and laboratory supplies, instruments, and equipments: $130,000.

Planting on national forests: For the purchase of tree seed, cones, and nursery stock, for seeding and tree planting within national forests, and for experiments and investigations necessary for such seeding and tree planting, $225,000;

Reconnaissance, national forests: For estimating and appraising timber and other resources on the national forests preliminary to disposal by sale or to the issue of occupancy permits, and for emergency expenses incident to their sale or use $121,000.

Improvement of the national forests: For the construction and maintenance of roads, trails, bridges, fire lanes, telephone lines, cabins, fences, and other improvements necessary for the proper and economical administration, protection and development of the national forests, $2,500,000, of which amount $150,000 is reserved for expenditure on the Angeles, Cleveland, Santa Barbara, and San Bernardino National Forests in Southern California: *Provided*, That such sum of $150,000 shall not be expended unless an equal amount is contributed for such work by State, county, municipal, and/or other local interests, to be paid, in whole or in part, in advance of the performance of the work for which this appropriation provides: *Provided further*, That where, in the opinion of the Secretary of Agriculture, direct purchase will be more economical than construction, telephone lines, cabins, fences and other improvements may be purchased: *Provided further,* That not to exceed $100,000 may be expended for the construction and maintenance of boundary and range division fences, counting corrals, stock driveways and bridges, the development of stock-watering places, and the eradication of poisonous plants on the national forests: *Provided further,* That not to exceed $1,000 of this appropriation may be used for repair and maintenance of the dam at Cass Lake, Minnesota: *Provided further,* That not less than $1,500,000 of this appropriation shall be available only for the construction and maintenance of roads and trails.

FOREST RESEARCH

For forest research in accordance with the provisions of sections 1, 2, 7, 8, 9, and 10 of the Act entitled "An Act to insure adequate supplies of timber and other forest products for the people of the United States, to promote the full use for timber growing and other purposes of forest lands in the United States, including farm wood lots and those abandoned areas not suitable for agricultural production, and to secure the correlation and the most economical conduct of forest research in the Department of Agriculture through research in reforestation, timber growing, protection, utilization, forest economics, and related subjects," approved May 22, 1928 (U. S. C., Supp. III, title 16, secs. 581, 581a, 581f-581i), as follows:

Forest management: Fire, silvicultural, and other forest investigations and experiments under section 2, at forest experiment stations or elsewhere, $488,500.

Range investigations: Investigations and experiments to develop improved methods of management of forest and other ranges under section 7, at forest or range experiment stations or elsewhere, $85,000.

Forest products: Experiments, investigations, and tests of forest products under section 8, at the Forest Products Laboratory, or elsewhere, $635,000.

For carrying out the provisions of the Act entitled "An Act to provide for the acceptance of a donation of land and the construction thereon of suitable buildings and appurtenances for the forest products laboratory, and for other purposes," approved April 15, 1930, $100,000; and in addition thereto the Secretary of Agriculture is authorized to enter into contracts or otherwise to incur obligations for the purposes of such Act in amounts not exceeding $800,000.

Forest survey: A comprehensive forest survey under section 9, $125,000.

Forest economics: Investigations in forest economics under section 10, $50,000.

In all, salaries and general expenses, $12,066,230; and in addition thereto there are hereby appropriated all moneys received as contributions toward coöperative work under the provisions of section 1 of the Act approved March 3, 1925 (U. S. C., title 16, sec. 572), which funds shall be covered into the Treasury and constitute a part of the special funds provided by the Act of June 30, 1914 (U. S. C., title 16, sec. 498) : *Provided,* That not to exceed $470,076 may be expended for departmental personal services in the District of Columbia.

FOREST-FIRE COÖPERATION

For coöperation with the various States or other appropriate agencies in forest-fire prevention and suppression and the protection of timbered and cut-over lands in accordance with the provisions of sections 1, 2, and 3 of the Act entitled "An Act to provide for the protection of forest lands, for the reforestation of denuded areas, for the extension of national forests, and for other purposes, in order to promote continuous production of timber on lands chiefly valuable therefor," approved June 7, 1924 (U. S. C., title 16, secs. 564-570), as amended, including also the study of the effect of tax laws and the investigation of timber insurance as provided in section 3 of said act, $1,700,000, of which $34,320 shall be available for departmental personal services in the District of Columbia and not to exceed $3,000 for the purchase of supplies and equipment required for the purposes of said Act in the District of Columbia.

COÖPERATIVE DISTRIBUTION OF FOREST PLANTING STOCK

For coöperation with the various States in the procurement, production, and distribution of forest-tree seeds and plants in establishing windbreaks, shelter belts, and farm wood lots upon denuded or non-forested lands within such coöperating States, under the provisions of section 4 of the Act entitled "An Act to provide for the protection of forest lands, for the reforestation of denuded areas, for the extension of national forests, and for other purposes, in order to promote the continuous production of timber on lands chiefly suitable therefor," approved June 7, 1924 (U. S. C., title 16, sec. 567), and Acts supplementary thereto, $93,000, of which amount not to exceed $1,840 may be expended for departmental personal services in the District of Columbia.

ACQUISITION OF ADDITIONAL FOREST LANDS

For the acquisition of additional lands under the provisions of the Act of March 1, 1911 (U. S. C., title 16, secs. 513-519), as amended by the Act of June 7, 1924 (U. S. C., title 16, secs. 564-570), subject to the provisions of the Act of April 30, 1928 (45 Stat., p. 468), $2,000,000, of which amount not to exceed $35,940 may be expended for departmental personal services and supplies and equipment in the District of Columbia.

Total, Forest Service, $15,859,230.

* * * *

INTERCHANGE OF APPROPRIATIONS

Not to exceed 10 per centum of the foregoing amounts for the miscellaneous expenses of the work of any bureau, division, or office herein provided for shall be available interchangeably for expenditures on the objects included within the general expenses of such bureau, division, or office, but no more than 10 per centum shall be added to any one item of appropriation except in cases of extraordinary emergency, and then only upon the written order of the Secretary of Agriculture: *Provided*, That a statement of any transfers of appropriations made hereunder shall be included in the annual Budget.

MISCELLANEOUS

* * * *

PASSENGER-CARRYING VEHICLES

That not to exceed $125,000 of the lump-sum appropriations herein made for the Department of Agriculture shall be available for the purchase of motor-propelled and horse-drawn passenger-carrying vehicles necessary in the conduct of the field work of the Department of Agriculture outside the District of Columbia: *Provided*, That such vehicles shall be used only for official service outside the District of Columbia, but this shall not prevent the continued use for official service of motor trucks in the District of Columbia: *Provided further*, That the Secretary of Agriculture is authorized to expend from the funds provided for carrying out the provisions of the Federal Highway Act of November 9, 1921 (U. S. C., title 23, secs. 21 and 23), not to exceed $40,000 for the purchase of motor-propelled passenger-carrying vehicles to replace such vehicles heretofore acquired and used by the Secretary of Agriculture in the construction and maintenance of national-forest roads or other roads constructed under his direct supervision which are or may become unserviceable, including the replacement of not to exceed two such vehicles for use in the administrative work of the Bureau of Public Roads in the District of Columbia: *Provided further*, That appropriations contained in this Act shall be available for the maintenance, operation, and repair of motor-propelled and horse-drawn passenger-carrying vehicles, but expenditures for that purpose, exclusive of garage rent, pay of operator, tires, fuel, and lubricants, on any one motor-propelled passenger-carrying vehicle except a bus, used by the Department of Agriculture shall not exceed one-third of the market price of a new vehicle of the same make or class, and in any case not more than $500: *Provided further*, That the Secretary of Agriculture may exchange motor-propelled and horse-drawn vehicles, tractors, road equipment, and boats, and parts, accessories, tires, or equipment thereof, in whole or in part payment for vehicles, tractors, road equipment, or boats, or parts, accessories, tires, or equipment of such vehicles, tractors, road equipment, or boats, purchased by him.

14

Mileage Rates for Motor Vehicles

Whenever, during the fiscal year ending June 30, 1931, the Secretary of Agriculture shall find that the expenses of travel and administration, including travel and administration at official stations, can be reduced thereby, he may, in lieu of actual operating expenses, under such regulations as he may prescribe, authorize the payment of not to exceed 3 cents per mile for motor cycle or 7 cents per mile for an automobile, used for necessary travel on official business: *Provided*, That the Secretary of Agriculture may authorize not to exceed 10 cents per mile for an automobile used in localities where poor road conditions or high cost of motor supplies prevail and he finds that the average cost to the operator is in excess of 7 cents per mile: *Provided further*, That the Secretary of Agriculture may authorize the payment of toll and ferry charges, storage and towage for such motor cycles and automobiles, in addition to mileage allowance.

* * * *

Forest Roads and Trails

For carrying out the provisions of section 23 of the Federal Highway Act approved November 9, 1921 (U. S. C., title 23, sec. 23), including not to exceed $53,563 for departmental personal services in the District of Columbia, $7,500,000, which sum is composed of $1,445,000, part of the sum of $7,500,000 authorized to be appropriated for the fiscal year 1930 by the Act approved May 26, 1928 (45 Stat., p. 750), and $6,055,000, part of the amount authorized to be appropriated for the fiscal year 1931 by the Act approved May 26, 1928; *Provided*, That the Secretary of Agriculture shall, upon the approval of this Act, apportion and prorate among the several States, Alaska, and Porto Rico, as provided in section 23 of said Federal Highway Act, the sum of $7,500,000 authorized to be appropriated for the fiscal year ending June 30, 1931, by the Act approved May 26, 1928: *Provided further*, That the Secretary of Agriculture shall incur obligations, approve projects, or enter into contracts under his apportionment and prorating of this authorization, and his action in so doing shall be deemed a contractual obligation on the part of the Federal Government for the payment of the cost thereof: *Provided further*, That the total expenditures on account of any State or Territory shall at no time exceed its authorized apportionment: *Provided further*, That this appropriation shall be available for the rental, purchase, or construction of buildings necessary for the storage of equipment and supplies used for road and trail construction and maintenance, but the total cost of any such building purchased or constructed under this authorization shall not exceed $1,500: *Provided further*, That there shall be available from this appropriation not to exceed $15,000 for the acquisition by purchase, condemnation, gift, grant, dedication, or otherwise of land

and not to exceed $120,000 for the acquisition by purchase or construction of a building or buildings for the storage and repair of Government equipment for use in the construction and maintenance of roads.

1930—Act of May 27, 1930 (Public No. 268, Seventy-first Congress)—An Act To facilitate and simplify national-forest administration.

[SECTION 1]. That the Secretary of Agriculture is authorized to expend not to exceed $8,000 annually, out of any money appropriated for the improvement or protection of the national forests, for the fiscal year 1930 or for subsequent years, in the completion of water supply or sanitary systems costing in excess of the $500 limitation as imposed by the Act of March 3, 1925 (Forty-third Statutes, page 1132).

SEC. 2. That the Secretary of Agriculture is authorized to reimburse owners of private property for damage or destruction thereof caused by employees of the United States in connection with the protection, administration, or improvement of the national forests, payment to be made from any funds appropriated for the protection, administration, and improvement of the national forests: *Provided,* That no payment in excess of $500 shall be made on any such claim.

SEC. 3. That the Secretary of Agriculture is authorized in cases of emergency to incur such expenses as may be necessary in searching for persons lost in the national forests and in transporting persons seriously ill, injured, or who die within the national forests to the nearest place where the sick or injured person, or the body, may be transferred to interested parties or local authorities.

1930—Act of June 2, 1930 (Public No. 298, Seventy-first Congress)—An Act Authorizing appropriations to be expended under the provisions of sections 4 to 14 of the Act of March 1, 1911, entitled "An Act to enable any State to coöperate with any other State or States, or with the United States, for the protection of the watersheds of navigable streams, and to appoint a commission for the acquisition of lands for the purpose of conserving the navigability of navigable rivers," as amended.

That there is hereby authorized to be appropriated, out of any money in the United States Treasury not otherwise appropriated, to be expended under the provisions of sections 4 to 14 of the Act of March 1, 1911 (U. S. C., title 16, secs. 513 to 521), as amended by the Acts of March 4, 1913 (U. S. C., title 16, sec. 518), June 30, 1914 (U. S. C., title 16, sec. 500), and June 7, 1924 (U. S. C., title 16, sec. 570), not to exceed $3,000,000 for the fiscal year beginning July 1, 1931, and not to exceed $3,000,000 for the fiscal year beginning July 1, 1932.

1930—Act of June 9, 1930 (Public No. 319, Seventy-first Congress)—An Act Authorizing the Secretary of Agriculture to enlarge tree-planting operations on national forests, and for other purposes.

[SECTION 1]. That the Secretary of Agriculture is hereby authorized to establish forest tree nurseries and do all other things needful in preparation for planting on national forests on the scale possible under the appropriations authorized by this Act: *Provided,* That nothing in this Act shall be deemed to restrict the authority of the said Secretary under other authority of law.

SEC. 2. There is hereby authorized to be appropriated for the fiscal year ending June 30, 1932, not to exceed $250,000; for the fiscal year ending June 30, 1933, not to exceed $300,000; for the fiscal year ending June 30, 1934, not to exceed $400,000; and for each fiscal year thereafter, not to exceed $400,000, to enable the Secretary of Agriculture to establish and operate nurseries, to collect or to purchase tree seed or young trees, to plant trees, and to do all other things necessary for reforestation by planting or seeding national forests and for the additional protection, care, and improvement of the resulting plantations or young growth.

SEC. 3. The Secretary of Agriculture may, when in his judgment such action will be in the public interest, require any purchaser of national-forest timber to make deposits of money, in addition to the payments for the timber, to cover the cost to the United States of (1) planting (including the production or purchase of young trees), (2) sowing with tree seeds (including the collection or purchase of such seeds), or (3) cutting, destroying, or otherwise removing undesirable trees or other growth, on the national-forest land cut over by the purchaser, in order to improve the future stand of timber: *Provided,* That the total amount so required to be deposited by any purchaser shall not exceed, on an acreage basis, the average cost of planting (including the production or purchase of young trees) other comparable national-forest lands during the previous three years. Such deposits shall be covered into the Treasury and shall constitute a special fund, which is hereby appropriated and made available until expended, to cover the cost to the United States of such tree planting, seed sowing, and forest improvement work, as the Secretary of Agriculture may direct: *Provided,* That any portion of any deposit found to be in excess of the cost of doing said work shall, upon the determination that it is so in excess, be transferred to miscellaneous receipts, forest reserve fund, as a national-forest receipt of the fiscal year in which such transfer is made: *Provided further,* That the

Secretary of Agriculture is authorized, upon application of the Secretary of the Interior, to furnish seedlings and/or young trees for replanting of burned-over areas in any national park.

1930—Act of June 23, 1930 (Public No. 412, Seventy-first Congress)—An Act To reorganize the Federal Power Commission.

[SECTION 1]. That sections 1 and 2 of the Federal Water Power Act are amended to read as follows:

" That a commission is hereby created and established, to be known as the Federal Power Commission (hereinafter referred to as the ' commission ') which shall be composed of five commissioners who shall be appointed by the President, by and with the advice and consent of the Senate, one of whom shall be designated by the President as chairman and shall be the principal executive officer of the commission: *Provided,* That after the expiration of the original term of the commissioner so designated as chairman by the President, chairmen shall be elected by the commission itself, each chairman when so elected to act as such until the expiration of his term of office.

" The commissioners first appointed under this section, as amended, shall continue in office for terms of one, two, three, four, and five years, respectively, from the date this section, as amended, takes effect, the term of each to be designated by the President at the time of nomination. Their successors shall be appointed each for a term of five years from the date of the expiration of the term for which his predecessor was appointed, except that any person appointed to fill a vacancy occurring prior to the expiration of the term of which his predecessor was appointed shall be appointed only for the unexpired term of such predecessor. Not more than three of the commissioners shall be appointed from the same political party. No person in the employ of or holding any official relation to any licensee or to any person, firm, association, or corporation engaged in the generation, transmission, distribution, or sale of power, or owning stock or bonds thereof, or who is in any manner pecuniarily interested therein, shall enter upon the duties of or hold the office of commissioner. Said commissioners shall not engage in any other business, vocation, or employment. No vacancy in the commission shall impair the right of the remaining commissioners to exercise all the powers of the commission. Three members of the commission shall constitute a quorum for the transaction of business, and the commission shall have an official seal of which judicial notice shall be taken. The commission shall annually elect a vice chairman to act in case of the absence or disability of the chairman or in case of a vacancy in the office of chairman.

" Each commissioner shall receive an annual salary of $10,000, together with necessary traveling and subsistence expenses, or per diem allowance in lieu thereof, within the limitations prescribed by law, while away from the seat of government upon official business.

" The principal office of the commission shall be in the District of Columbia, where its general sessions shall be held; but whenever the convenience of the public or of the parties may be promoted or delay or expense prevented thereby, the commission may hold special sessions in any part of the United States.

" Sec. 2. The commission shall have authority to appoint, prescribe the duties, and fix the salaries of, a secretary, a chief engineer, a general counsel, a solicitor, and a chief accountant; and may, subject to the civil service laws, appoint such other officers and employees as are necessary in the execution of its functions and fix their salaries in accordance with the Classification Act of 1923, as amended. The commission may request the President to detail an officer or officers from the Corps of Engineers, or other branches of the United States Army, to serve the commission as engineer officer or officers, or in any other capacity, in field work outside the seat of government, their duties to be prescribed by the commission; and such detail is hereby authorized. The President may also, at the request of the commission, detail, assign, or transfer to the commission engineers in or under the Departments of the Interior or Agriculture for field work outside the seat of government under the direction of the commission.

" The commission may make such expenditures (including expenditures for rent and personal services at the seat of government and elsewhere, for law books, periodicals, and books of reference, and for printing and binding) as are necessary to execute its functions. Expenditures by the commission shall be allowed and paid upon the presentation of itemized vouchers therefor, approved by the chairman of the commission or by such other member or officer as may be authorized by the commission for that purpose."

Sec. 2. Subsection (c) of section 4 of the Federal Water Power Act is amended by adding at the end thereof the following new sentence: " Such report shall contain the names and show the compensation of the persons employed by the commission."

Sec. 3. Notwithstanding the provisions of section 1 of this Act the Federal Power Commission as constituted upon the date of the approval of this Act shall continue to function until the date of the reorganization of the commission pursuant to the provisions of such section. The commission shall be deemed to be reorganized upon such date as three of the commissioners appointed as provided in such section 1 have taken office, and no such commissioner shall be paid salary for any period prior to such date.

Sec. 4. This Act shall be held to reorganize the Federal Power Commission created by the Federal Water Power Act, and said Federal Water Power Act shall remain in full force and effect, as herein amended, and no regulations, actions, investigations, or other proceedings under the Federal Water Power Act existing or pending at the time of the approval of this Act shall abate or otherwise be affected by reasons of the provisions of this Act.

1930—Act of July 3, 1930 (Public No. 519, Seventy-first Congress)—An Act Making appropriations to supply deficiencies in certain appropriations for the fiscal year ending June 30, 1930, and prior fiscal years, to provide supplemental appropriations for the fiscal years ending June 30, 1930, and June 30, 1931, and for other purposes.

* * * *

DEPARTMENT OF AGRICULTURE

* * * *

FOREST SERVICE

The unexpended balance of the appropriation of $35,000, contained in the first deficiency Act, fiscal year 1930, approved March 26, 1930, for carrying into effect the provisions of the Act entitled "An Act to authorize the improvement of the Oregon Caves, in the Siskiyou National Forest," approved February 28, 1929 (45 Stat., p. 1407), is hereby continued available for the same purposes until June 30, 1931.

* * * *

MISCELLANEOUS

Forest roads and trails: For an additional amount for carrying out the provisions of section 23 of the Federal Highway Act approved November 9, 1921, including the same objects specified under this head in the Agricultural Appropriation Act for the fiscal year 1931, and including not to exceed $24,500 for departmental personal services in the District of Columbia, $3,500,000, which sum is composed of $1,445,000, part of the sum of $7,500,000 authorized to be appropriated for the fiscal year 1931 by the Act approved May 26, 1928, and $2,055,000, part of the sum of $5,000,000 authorized to be appropriated for the fiscal year 1931, by the Act approved May 5, 1930: *Provided,* That the Secretary of Agriculture shall, upon approval of this Act, apportion and prorate among the several States, Alaska, and Porto Rico, as provided in section 23 of the said Federal Highway Act, the sum of $5,000,000 authorized to be appropriated for the fiscal year ending June 30, 1931, by the Act approved May 5, 1930: *Provided further,* That the Secretary of Agriculture shall incur obligations, approve projects, or enter into contracts under his apportionment and prorating of this authorization, and his action in so doing shall be deemed a contractual obligation on the part of the Federal Government for the payment of the cost thereof: *Provided further,* That the total expenditures on account of any State or Territory shall at no time exceed its authorized apportionment.

* * * *

APPENDIX 5

FINANCIAL STATEMENTS

Explanatory Note

Statements showing appropriations, receipts, expenditures and other financial data for a series of years constitute the most effective single means of exhibiting the growth and development of a service. Due to the fact that Congress has adopted no uniform plan of appropriation for the several services and that the latter employ no uniform plan in respect to the recording and reporting of their receipts and expenditures, it is impossible to present data of this character according to any standard scheme of presentation. In the case of some services the administrative reports contain tables showing financial conditions and operations of the service in considerable detail; in others financial data are almost wholly lacking. Careful study has in all cases been made of such data as are available, and the effort has been made to present the results in such a form as will exhibit the financial operations of the services in the most effective way that circumstances permit.

From 1877 to 1880 forestry work in the Department of Agriculture was on an allotment basis, $2000 being made available for 1877 and $2500 for each of the three succeeding years. The Division of Forestry began in 1881 with an appropriation of $5000, and continued on that basis for two years. In 1883 it received $10,000, and the same amount annually through 1890. From 1891· to 1910 the record shows:

Division of Forestry: Appropriations, Fiscal Years 1891 to 1900

Year	Salaries	General expenses	Totals
1891	$7,820	$10,000	$17,820
1892	7,820	15,057	22,877
1893	7,820	12,000	19,820
1894	7,820	20,000	27,820
1895	8,320	20,000	28,320
1896	8,520	25,000	33,520
1897	8,520	20,000	28,520
1898	8,520	20,000	28,520
1899	8,520	20,000	28,520
1900	8,520	40,000	48,520

Bureau of Forestry: Appropriations, Fiscal Years 1901 to 1905

Year	Salaries	General expenses	Totals
1901	$8,520	$80,000	$88,520
1902	39,160	146,280	185,440
1903	37,860	254,000	291,860
1904	37,140	312,860	350,000
1905	37,140	402,733	439,873

Forest Service: General Appropriations, Fiscal Years, 1906 to 1918

Year	Salaries	General expenses	Improvements	Totals
1906	$81,960	$1,113,258	......	$1,195,218
1907	112,860	1,795,469	......	1,908,329
1908	143,200	3,429,722	......	3,572,922
1909	144,300	3,151,900	$600,000	3,896,200
1910	60,200	3,986,000	600,000	4,646,200
1911	60,200	5,602,900	275,000	5,938,100
1912	2,318,680	3,714,420 [a]	500,000	6,533,100
1913	2,235,760	2,907,285 [b]	400,000	5,543,045
1914	2,239,560	2,960,119 [c]	400,000	5,599,679
1915	2,305,160	3,292,339	400,000	5,997,499
1916	2,335,580	2,874,976	400,000	5,610,556
1917	2,358,505	2,788,415	400,000	5,546,920
1918	2,632,885 [d]	3,589,475	450,000	6,672,360

[a] Includes $1,000,000 to be used for fire fighting in case of emergency. Of this, only $46,039 was used.

[b] Includes $200,000 to be used for fire fighting in case of emergency. Of this, only $47,777 was used.

[c] Includes $200,000 to be used for fire fighting in case of emergency. Of this, only $41.60 was used.

[d] Includes $185,085, " Increase of Compensation."

SPECIAL APPROPRIATIONS, 1907 TO 1918

	1907	1908	1909	1910	1911	1912	1913
Wichita Forest and Game Preserve	$15,000	......	......	......	......	......	......
Survey and report on Appalachian and White Mountain Watersheds	......	$25,000	......	......	......	......	......
Naval stores industry	......	......	$10,000	......	......	......	......
Acquisition of forest lands (Act Mar. 1, 1911)	......	......	......	a $1,000,000	$2,000,000	$2,000,000	$2,000,000
Forest fire coöperation (Act Mar. 1, 1911)	......	......	......	......	200,000	......	......
National Forest Reservation Commission (Act Mar. 1, 1911)	......	......	......	......	25,000	25,000	25,000
Coöperative work in forest investigations (Act June 30, 1906)	21,410	26,629	40,779	29,164	6,860	3,221	6,748
Payments to states for schools and roads (Acts June 30, 1906, and May 23, 1908)	75,781	163,082	447,064	441,522	510,091	484,639	518,292
Payments to Arizona and New Mexico school funds (Act June 20, 1910)	......	......	......	......	815	30,434	36,088
Roads and trails for states (Act Aug. 10, 1912)	......	......	25,000	......	......	......	207,305

a Not used.

	1914	1915	1916	1917	1918
Acquisition of forest lands (Act Mar. 1, 1911)	$2,000,000	$2,000,000	$100,000	$1,000,000	$2,000,000
Forest fire coöperation (Act Mar. 1, 1911)	75,000	100,000	100,000	100,000	100,000
National Forest Reservation Commission (Act Mar. 1, 1911)	25,000	25,000	25,000	25,000	25,000
Coöperative work in forest investigations (Acts of June 30, 1906, June 30, 1914, and Aug. 11, 1916)	10,321	10,863	58,401	106,111	183,269
Payments to states for schools and roads (Act May 23, 1908)	586,593	599,272	610,789	695,541	848,874
Payments to Arizona and New Mexico for schools (Act June 20, 1910)	45,547	40,622	38,278	41,375	61,532
Roads and trails for states (Act Mar. 4, 1913)	234,639	239,709	244,319	278,217	389,550
Land exchange, State of Montana (Act Mar. 4, 1913)	25,000	......	......	......	......
Land exchange, State of Washington (Act Mar. 4, 1915)	......	......	50,000	......	......
Protection of Oregon and California R. R. lands (transfer from Interior)	......	25,000	25,000	25,000	35,000
Coöperative construction of roads and trails (Act July 11, 1916)	......	......	......	1,000,000	1,000,000
Coöperative construction of bridge, Minnesota National Forest	......	......	......	......	10,000

APPROPRIATIONS, FISCAL YEARS 1919 TO 1926

General Fund	1926	1925	1924	1923	1922	1921	1920	1919
Salaries	$3,325,003	$3,325,003	$2,471,000	$2,465,020	$2,465,020	[a] $2,468,380	$2,485,660	$2,488,620
Increase of compensation (bonus)	2,977,973	2,888,923	732,227	746,302	762,478	717,796	590,854	239,606
Administration, national forests	[c] 1,058,000	[b] 1,590,000	2,461,862	2,485,362	2,412,362	2,225,262	2,069,201	1,803,127
Fire fighting and prevention	50,000		399,000	625,000	591,000	1,025,000	3,100,000	815,000
Airplane fire control					50,000			
Insect control	25,000	25,000	31,000	60,000	75,000	87,000	107,000	119,800
Land selection	60,900	60,900	60,000	10,000				
Camp improvements	25,000	25,000	15,000		150,000	150,000	161,100	161,100
Supplies and equipment	140,480	180,480	150,000	150,000	120,640	120,640	145,640	145,640
Tree planting	131,705	131,705	125,640	125,640	100,000	80,000	80,000	100,000
Timber and grazing reconnaissance	108,550	108,550	100,000	100,000				
Improvements	[d] 431,900	426,900	448,000	425,000	400,000	400,000	450,000	450,000
Silvicultural experiments	202,020	202,020	135,000	85,000	85,000	50,000	78,728	78,728
Range investigations	40,320	40,320	35,000	35,000	35,000	35,000	35,000	35,000
Forest products investigations	383,264	383,264	350,800	340,000	325,000	223,260	173,260	173,260
Miscellaneous investigations	33,800	33,800	31,280	31,280	31,280	31,280	31,280	31,280
Total, Salaries and General Expenses	$8,998,915	$9,376,865	$7,545,809	$7,683,604	$7,602,780	$7,663,618	$9,507,723	$6,636,161
Forest fire coöperation (Acts Mar. 1, 1911, and June 7, 1924)	$660,000	$401,900	$400,000	$400,000	$400,000	$125,000	$100,000	$100,000
Acquisition of forest lands (Act Mar. 1, 1911)	1,000,000	818,540	450,000	450,000	1,000,000		600,000	
National Forest Reservation Commission (Act Mar. 1, 1911)	25,000	25,000	25,000	25,000	25,000	25,000	25,000	25,000
Protection of Oregon and California R. R. lands (Act Mar. 1, 1911)			54,600	51,480	[e] 40,896	[e] 36,512	[e] 39,110	[e] 25,000
Aircraft production research (transfer from War)						50,000	25,000	221,000
Box and packing investigations (transfer from War)							15,000	35,234

[a] A misprint in the appropriation act shows this as $2,478,380.　　[b] After transfer of $8,000 to War Dept.　　[c] $100,000 available in 1925.
[d] $5,000 available in 1927.　　[e] Transfer from Interior.

APPROPRIATIONS, FISCAL YEARS 1919 TO 1926—Continued

General fund	1919	1920	1921	1922	1923	1924	1925	1926
Aircraft production research (transfer from Navy)	$118,000	......	......	......	......	......	......	......
Aviation (transfer from Navy)	......	......	$80,000	$50,000	......	......	$28,250	$26,400
Construction and repair (transfer from Navy)	......	......	......	......	......	$55,000	2,500	......
Air service (transfer from War)	......	......	......	15,000	$10,000	1,000	10,000	3,000
Forest products tests (transfer from War)	......	......	4,500	......	......	10,000	......	......
Forest insect control (transfer from Interior)	......	......	......	......	1,000	......	......	......
Insect control, Oregon and California (available to Dec. 31, 1926)	......	......	100,000	150,000	......	......	......	......
Olympic National Forest, emergency	......	......	......	......	33,000	25,000	......	......
Motor boats, Alaska	......	......	......	......	8,500	......	......	......
Coöperative distribution of planting stock (Act of June 7, 1924)	......	......	......	......	......	......	......	50,000
Special Funds								
Coöperative work (Acts of June 30, 1906, June 30, 1914, and Aug. 11, 1916)	$491,854	$2,044,592	$1,946,041	$1,394,827	$1,514,772	$2,618,442	$2,104,219	$1,920,222
Payments to states for school and roads (Act May 23, 1908)	876,334	1,069,887	1,180,063	1,023,084	846,442	1,821,423	1,801,848	1,242,954
Payments to Arizona and New Mexico for schools (Act June 20, 1910)	69,593	78,867	73,280	59,596	35,762	50,127	44,504	28,322
Roads and trails for states (Act Mar. 4, 1913)	350,534	427,954	472,025	409,234	338,577	528,569	520,739	497,182
Road Funds								
Coöperative construction of roads and trails (Act July 11, 1916)	$1,000,000	$1,000,000	$1,000,000	$1,000,000	$1,000,000	$1,000,000	$1,000,000	$1,000,000
Federal Forest Road Construction (Act Feb. 28, 1919)	3,000,000	3,000,000	3,000,000	......	......	......	......	......
Forest Road Development (Act Nov. 9, 1921)	......	......	......	2,500,000	3,000,000	3,000,000	3,000,000	3,000,000
Forest Highways (Act Nov. 9, 1921)	......	......	......	2,500,000	7,000,000	3,500,000	3,500,000	4,500,000

APPROPRIATIONS AND EXPENDITURES, FISCAL YEARS 1927 TO 1930

General fund	1927		1928		1929		1930
	Appropriation	Expenditures	Appropriation	Expenditures	Appropriation	Expenditures	Appropriation
Salaries	$3,325,000.00	$3,323,686.00	$6,383,240.00	$6,379,653.12	$6,089,573.00	$6,911,546.42	$6,990,230.00
Administration, national forests	3,007,973.00	2,996,735.16	......	......	......	......	......
Fire fighting and prevention	a 2,380,000.00	2,370,626.89	b 927,000.00	c 1,005,962.30	d 1,225,000.00	1,218,613.20	e 100,000.00
Airplane fire patrol	50,000.00	46,563.48	50,000.00	47,382.28	50,000.00	43,949.37	50,000.00
Insect control	58,000.00	46,148.05	125,000.00	124,756.67	75,000.00	72,435.21	75,000.00
Land selection	55,000.00	53,659.45	55,000.00	52,575.58	52,546.00	50,420.27	52,500.00
Camp improvements	40,000.00	38,882.20	40,000.00	39,334.51	40,000.00	39,263.66	50,000.00
Supplies and equipment	130,000.00	128,200.79	180,000.00	129,587.10	180,100.00	129,410.00	180,000.00
Tree planting	131,700.00	131,184.35	150,000.00	148,888.85	212,220.00	211,037.44	210,000.00
Timber and grazing reconnaissance	108,550.00	108,093.26	f 107,800.00	107,570.04	g 110,470.37	110,861.17	108,550.00
Improvements	526,900.00	524,467.96	526,900.00	526,181.48	588,912.00	586,163.60	645,000.00
Silvicultural experiments	250,000.00	249,539.42	337,000.00	336,783.28	377,407.00	377,686.65	413,000.00
Range investigations	40,320.00	39,988.93	44,880.00	44,876.31	52,680.00	52,449.22	67,000.00
Forest products investigations	403,264.00	402,154.52	500,000.00	499,213.94	542,596.00	541,887.91	585,000.00
Miscellaneous investigations	33,800.00	33,670.01	......	......	......	......	......
Forest survey	......	......	......	......	......	......	40,000.00
Economic investigations	......	......	......	......	......	......	25,000.00
Total, Salaries and General Expenses.	$10,540,507.00	$10,493,400.47	$9,376,820.00	$9,442,768.46	$10,446,504.37	$10,345,723.12	$9,541,280.00
Forest fire coöperation (Act June 7, 1924)	$710,000.00	$703,885.48	$1,000,000.00	$991,700.83	$1,209,802.00	$1,208,024.31	$1,400,000.00
Acquisition of forest lands (Act Mar. 1, 1911)	1,000,000.00	1,000,000.00	2,000,000.00	2,000,000.00	1,000,000.00	1,000,000.00	2,000,000.00
National Forest Reservation Commission (Act Mar. 1, 1911)	25,000.00	867.94	25,000.00	351.71	25,000.00	500.00	25,000.00

a $150,000 available in 1926.   b $150,000 available in 1927.   c Includes $80,003.43 of 1929 appropriation.   d $100,000 available in 1928.   e Available in 1929.   f After transfer of $750 to Geological Survey.   g After transfer of $1,121.68 to Geological Survey.

APPROPRIATIONS AND EXPENDITURES, FISCAL YEARS 1927 TO 1930—*Continued*

General fund	1927		1928		1929		1930
	Appropriation	Expenditures	Appropriation	Expenditures	Appropriation	Expenditures	Appropriation
Coöperative distribution of forest planting stock (Act June 7, 1924)	$75,000.00	$71,466.96	$75,000.00	$74,977.78	$75,480.00	$75,187.59	$88,000.00
Advisory committee for aeronautics (transfer from Committee on Aeronautics)	2,500.00	2,500.00	2,500.00	2,500.00	2,500.00	2,499.27	......
Aviation (transfer from Navy)	23,700.00	23,700.00	11,500.00	11,500.00	5,000.00	5,000.00	......
Special Funds							
Coöperative work (Acts of June 30, 1906, June 30, 1914, Aug. 11, 1916, and Mar. 3, 1925) [h]	$1,507,463.34	$1,461,221.61	$1,550,816.42	$1,511,887.81	$1,880,457.86	$1,815,219.37	......
Payments to states for schools and roads (Act May 23, 1908)	1,285,523.44	1,285,523.44	1,285,101.70	1,285,101.70	1,351,436.52	1,351,436.52	......
Payments to Arizona and New Mexico for schools (Act June 20, 1910)	13,567.25	13,567.25	26,198.94	26,198.94	35,854.66	35,854.66	......
Roads and trails for states (Act Mar. 4, 1913) [h]	514,209.38	447,852.80	514,086.78	666,704.79	540,528.50	419,179.32	......
Refunds, excess deposits (Act Mar. 4, 1907)	60,396.91	60,396.91	124,919.58	124,919.58	81,176.35	81,176.35	......
Road Funds							
Forest road development (Act Nov. 9, 1921) [h]	$3,000,000.00	$3,531,934.57	$3,000,000.00	$3,928,179.03	$3,000,000.00	$3,801,944.98	$3,000,000.00
Forest highways (Act Nov. 9, 1921) [h]	4,500,000.00	4,949,201.17	4,500,000.00	4,504,864.96	4,500,000.00	5,100,359.88	4,500,000.00

[h] Balances at the close of each fiscal year are available for expenditure in subsequent years, and expenditures in one year are in some instances in excess of the appropriation for that year.

Net receipts from National Forests for the fiscal years 1905 to 1929 inclusive, and amounts paid to states and territories and transferred to the road and trail fund

Fiscal year	Net receipts	Paid school fund Arizona and New Mexico, Act June 20, 1910	Paid states and territories, Acts June 30, 1906, and May 23, 1908	Transferred to Road and Trail Fund, Acts Aug. 10, 1912, and Mar. 4, 1913
1905.....	73,276.15	......	......	......
1906.....	757,813.01	......	$75,781.33	......
1907.....	1,530,321.88	......	153,032.19	......
1908.....	1,788,255.19	......	447,063.78	......
1909.....	1,766,088.46	......	441,522.05	......
1910.....	2,041,181.22	$815.48	510,091.40	......
1911.....	1,968,993.42	30,434.16	484,639.24	......
1912.....	2,109,256.91	36,087.75	518,292.28	$207,304.66
1913.....	2,391,920.85	45,547.30	586,593.39	234,638.68
1914.....	2,437,710.21	40,621.52	599,272.17	239,708.86
1915.....	2,481,469.35	38,278.29	610,788.49	244,319.10
1916.....	2,823,540.71	41,375.12	695,541.40	278,216.56
1917.....	3,457,028.41	61,532.36	848,874.01	339,549.61
1918.....	3,574,930.07	69,592.59	876,334.39	350,533.75
1919.....	4,358,414.86	78,867.32	1,069,886.88	427,954.75
1920.....	4,793,482.28	73,229.75	1,180,063.13	472,025.25
1921.....	4,151,931.42	59,596.18	1,023,083.81	409,233.53
1922.....	3,421,531.22	35,761.60	846,442.41	338,576.96
1923.....	5,335,818.13	50,127.49	1,321,422.66	528,569.06
1924.....	5,251,903.11	44,503.87	1,301,848.22	520,739.29
1925.....	5,000,137.49	28,321.76	1,242,953.93	497,181.57
1926.....	5,155,661.02	13,567.25	1,285,523.44	514,209.38
1927.....	5,166,605.74	26,198.94	1,285,101.70	514,086.78
1928.....	5,441,600.72	35,854.66	1,351,436.52	540,528.50
1929.....	6,299,801.86	41,242.91 [a]	1,564,639.74 [a]	625,855.89 [b]
Totals...	$83,578,673.69	$851,556.30	$20,320,228.56	$7,283,232.18

[a] Not yet paid.   [b] Not yet transferred.

APPENDIX 6

BIBLIOGRAPHY [1]

Explanatory Note

The bibliographies appended to the several monographs aim to list only those works which deal directly with the services to which they relate, their history, activities, organization, methods of business, problems, etc. They are intended primarily to meet the needs of those persons who desire to make a further study of the services from an administrative standpoint. They thus do not include the titles of publications of the services themselves, except in so far as they treat of the services, their work and problems. Nor do they include books or articles dealing merely with technical features other than administrative of the work of the services. In a few cases explanatory notes have been appended where it was thought they would aid in making known the character or value of the publication to which they relate.

After the completion of the series the bibliographies may be assembled and separately published as a bibliography of the Administrative Branch of the National Government.

General

Aiding cities and towns to name their trees; the Forest service will identify trees in streets and parks. (In Plant world, June 1906, v. 9: 142-43).

Barnes, W. C. United States Forest service. (In Outwest, Aug. 1908, v. 29, no. 2: 89-109).

Black, R. L. Forest service. (In Outlook, Aug. 26, 1905, v. 80: 1020-28).

___

[1] Compiled by Sophy H. Powell.

Blanchard, C. J. Mutual relations of the Forest service and the Reclamation service. (In Forestry and irrigation, Jan. 1906, v. 12:42-43).

Brown, N. C. American lumber industry. . . . New York, John Wiley and sons, 1923. 279 p.

Forest service, p. 261-62.

Burdon, E. R. Report on a visit to the United States and Canada for the purpose of studying . . . forestry departments. . . . Cambridge, University press, 1912. 24 p.

Cook, A. S. Guardians of our forests. (In American forestry, Sept. 1919, v. 25:1329-33)

Cooper, A. W. Work of the Forest service. (In Water and forest, Jan. 1907, v. 6, no. 4:9-11)

DeBoer, S. R. Against transfer; Forest service should remain where it is. (In Parks and recreation, Jan.-Feb. 1922, v. 5, no. 3:227-30)

Denver, Colorado. Chamber of commerce and Board of trade. Committee on forestry. Report [with resolutions of commendation of U. S. Forest service]. Denver, no publisher, 1909. 13 p.

DuBois, Coert. Mountain communities and the Forest service. (In University of California journal of agriculture, Nov. 1916, v. 4, no. 3:71-74)

Forest service [Editorials]. (In Outlook, March 11–Apr. 22, 1925, v. 139:361-62, 404-05, 444-46, 485-86, 523, 565-66, 604-05)

Criticism of Forest service.

Graves, H. S. Farmer and the forester. (In Purdue agriculturist, Jan. 1917, v. 11:11-13, 52-54)

—— Forest service as an organization for national defence. (In Forestry Kaimin, Journal of school of forestry, Montana state university, 1917, v. 3:52-53)

—— New menace to forestry; reorganization of Forest service under proposed department of public works. (In American forestry, Oct. 1921, v. 27:645-47)

—— What the government is doing in forestry. (In Paper, Oct. 16, 1912, v. 9, no. 5:15-16, 38)

Ise, John. United States forest police. New Haven, Yale university press, 1920. 395 p.

Hostility to national forests, Chapters 8-9; work of Forest service, Chapter 10.

Ivy, T. P. Forestry problems in the United States. Hendersonville, N. C. 1906. 47 p.

Lamb, G. N. Work of the Forest service. (In Alpha Zeta quarterly, March, 1915, v. 13, no. 4:34-41)

Levison, J. J. Mission and scope of Forest service and its relation to the cooperage business. . . . (In Packages, Nov. 1905, v. 8, no. 11: 41-42)

Old order changes. (In Journal of forestry, Mar. 1920, v. 18:203-10)

Duties of Chief forester.

Pinchot, Gifford. Federal forest service. (In American forest congress, Proceedings, 1905: 390-99)

Potter, A. F. Forest service and the meat supply. (In Southwest stockman-farmer, Jan. 1, 1915, v. 31, no. 5:3)

—— Keep the Forest service in the Department of agriculture. (In Producer, Oct. 1921, v. 3, no. 5:9-10)

Price, O. W. Government work in forestry. . . . (In Woodcraft, Dec. 1905, v. 4:135-39)

—— Work of the Bureau of forestry. (In American forest congress, Proceedings, 1905: 355-63)

Shall the Forest service be eliminated from Alaska? (In American forestry, Jan. 1922, v. 28:37)

Stockbridge, H. E. Growth of the Forest service library. (In Forestry quarterly, June, 1910, v. 8: 198-200)

U. S. Congress. Joint committee to investigate Interior department and Forestry service. Compilation of letters, telegrams, reports and other documents offered in evidence . . . in the course of hearings held by committee Jan. 26–May 28, 1910. 1910. 2 v.

—— —— —— Investigation of Department of Interior and of Bureau of forestry. . . . 1914. 13 v. (61st Cong. Senate doc. 719)   Serials 5892-5903

—— —— House. Committee on agriculture. Agricultural appropriation bill. Hearings, 1904-1922, 1902-1921.

See Forest service in index to each volume.

—— —— —— —— Forestry. Hearings . . . , on H. R. 15327, Jan. 26 and 27, 1921. 1921. 58 p.

—— —— —— —— Forestry. Hearings . . . Jan. 9, 10, 11, 12, 1922. 1922. 266 p.

—— —— —— *Committee on appropriations.* Agricultural appropriation bills, 1923-date.

Hearing . . . 1922-date.

See Forest service in index.

Hearings for previous years were held before House committee on agriculture.

—— —— *Senate. Committee on agriculture and forestry.* Agricultural appropriation bill, 1905-1922. Hearings . . . 1904-1921.

See Forest service in index to each volume.

—— —— —— *Committee on appropriations.* Agricultural appropriation bill, 1923-date. 1922-date.

See Forest service in index to each annual volume.

United States Department of agriculture, Forest service. (In American institute of architects, Journal, July 14, 1917, v. 5 : 246-48)

U. S. Forest service. Annual report of the forester, 1883-date. 1887-date.

The first three reports were printed only in the reports of the Department of agriculture.

—— —— Division of forestry, by B. E. Fernow. (In United States Agriculture dept. Yearbook, 1897 : 143-60)

—— —— Forest service; what it is and how it deals with forest problems. 3d ed. 1907. 38 p.

(United States Forest service circ. 36)

—— —— Forest worker, Bi-monthly, Sept. 1924-date.

Mimeographed. Much the same material as appears in the Service bulletin in amplified form for general reader. Evident object is furtherance of coöperative features of Clarke-McNary act.

—— —— Forests and forestry in the United States, supplementing exhibit of the United States Forest service at Brazil centennial exposition Rio de Janeiro, 1922-23, by H. A. Smith. 1922. 16 p.

—— —— Government forest work. 1915. 19 p.
—— —— Same. 1916. 16 p.
—— —— Same. 1922. 47 p.
—— —— Same. 1924. 44 p.

—— —— Same. 1925. 44 p.
—— —— Same. 1927. 44 p.

(United States Agriculture dept. Circ. 211)

—— —— Service bulletin. Weekly, Nov. 1917-date.

Mimeographed for Service officers and employees only. Disseminates Service news and discusses problems, and thereby throws light on Service activities and organization.

Value of the United States Forest service. (In National geographic magazine, Jan. 1909, v. 20:29-41)

What the Forest service does. (In Outlook, Apr. 12, 1922, v. 130:586-88)

Whitney, L. Forest service of the United States and its work. (In School science and mathematics, Dec. 1910, v. 10: 814-19)

HISTORY

American association for the advancement of science. Message from the President. . . . Memorial upon cultivation of timber and preservation of forests. . . . 1874. 6 p. (43d Cong., 1st sess. Senate Ex. doc. 28)   Serial 1580

Barnes, W. C. Gifford Pinchot, forester. (In McClures' magazine, July 1908, v. 31:319-27)

Contains information as to development of Forest service.

Bryan, J. W. Proposed investigation of the Forest service . . . [speech in Congress]. Washington, no publisher, 1913. 32 p.

Early history of the Forest service.

Decade of progress in the Forest service. (In American forestry, Mar. 1920, v. 26:131-32).

Decennial celebration of Forest products laboratory. (In Chemical and metallurgical engineering, Aug. 18, 1920, v. 23:270-72)

Fernow, B. E. Economics of forestry. . . . New York, T. Y. Crowell and company, 1902. 520 p.

Forestry movement in the United States, p. 369-411.

—— Need of a forest administration for the United States. . . . Salem, Mass., Salem press publishing and printing company, 1889. 359-366 p.

—— Practicability of an American forest administration. Baltimore, American economic association, Publications, 1891, v. 6:259-285.

—— Report upon the forestry investigations of the U. S. Department of agriculture, 1877-98. 1899. 401 p. (55th Cong., 3d sess. House doc. 181)

—— Situation [what has been accomplished in forestry as a policy, a science, and an art]. (In Journal of forestry, Jan. 1917, v. 15:3-14)

Forest service in war times. (In Science, Feb. 7, 1919, n s. v. 49:141-42)

Graves, H. S. Federal forestry. (In Science, Nov. 28, 1913, n. s. v. 38:753-58)

Greeley, W. B. Accomplishment of the Forest service in silviculture during the past decade, Feb. 15, 1915. 21 p.

Typewritten.

—— National forestry. (In Cornell university, Proceedings at opening of forestry building, May 15, 1914:5-15)

Hough, F. B. Report upon forestry. From committee appointed to memorialize Congress and the State legislature regarding cultivation of timber and preservation of forests. Salem, Mass., Salem press. 1878. 14 p.

Kinney, J. P. Forest legislation in America prior to Mar. 4, 1789. Ithaca, N. Y. (Cornell university, Agricultural experiment station of New York Bulletin 370, Jan. 1, 1916:358-405)

National academy of sciences. Report of committee upon inauguration of a forestry policy to the Secretary of interior. 1897. 47. p.

Pinchot, Gifford. Government forestry abroad. Baltimore, American economic association, Publications, 1891, v. 6:185-238.

Sherman, E. A. Thirty-five years of national forest growth. (In Journal of forestry, Feb. 1926, v. 24:129-35)

Stahl, R. M. Ballinger-Pinchot controversy. Northampton, Mass. Smith College Dept. of history, 1926. 69-138 p. (Smith College Studies in history, vol. XI, no. 2, Jan. 1926)

Sudworth, G. B. Origin and development of forest work in the United States (In Michigan political science association, Publications, Jan. 1902, v. 4:376-405)

Swain, E. H. F. Australian study of American forestry. . . . Brisbane, A. J. Cumming, government printer, 1918. 138 p.

Toumey, J. W. Recent progress and trends in forestry in the United States. (In Journal of forestry, Jan. 1925, v. 23:1-9)

U. S. Agriculture dept. Brief statutory history of the U. S. Department of agriculture, by F. G. Caffey. 1916. 26 p.
Forest service, 9-13

—— —— Historical sketch of the U. S. Department of agriculture, compiled by C. H. Greathouse. 1907. 97 p.

—— —— Summary of war work of Department of agriculture from April, 1917–June 30, 1919. 1919. 16 p.
Mimeographed.
Forest service, p. 2-4.

—— *Congress. Joint committee to investigate Interior department and Forestry service.* Investigation of Department of interior and Bureau of forestry. Hearings, . . . 1910. 7 v.

—— —— *House. Committee on public lands.* Cultivation of timber and the preservation of forests, 1874. 119 p. (43d Cong., 1st sess. House rept. 259).   Serial 1623

Wanlass, W. L. United States Department of agriculture, a study in administration. Baltimore, Johns Hopkins press, 1920. 132 p.
See Forest service in index.

Wartime work of the Forest service. (In Science, Dec. 28, 1917, n. s. v. 46: 632-33)

General Activities and Policies

American forest congress. Proceedings, 1905. . . . Washington, D. C. Suter publishing company, 1905. 474 p.
National and state forest policy, p. 355-451.

American paper and pulp association. Committee on forest conservation. Suggestions for a national forest policy. . . . New York? no publisher, 1919. 8 p.

Beaman, D. C. National forests and the Forest service. Albuquerque, N. M. no publisher, 1908. 20 p.
Attack on Forest service.

Berry, J. B. Farm woodland. Yonkers-on-Hudson, N. Y. World book company, 1923. 425 p.
Forest situation in the United States, p. 91-120

Colonel Graves on the Snell bill. (In American forestry, Mar. 1922, v. 28: 179-80)

Cox, W. T. and Cornwall, G. M. Policy of Forest service in selling timber and fixing stumpage prices. (In American lumberman, Aug. 21, 1909, no. 1787: 35-36)

Cunningham, R. N. Clarke-McNary law program. (In Idaho forester, annual edition, 1926, v. 8: 14-15)

Defebaugh, J. E. Appalachian forest reserves; a consideration of some features of the forest reserve policy of the United States. Chicago, Association of commerce, 1908. 13 p.

Fight for Alaska's forests. (In American forestry, April, 1922, v. 28: 201-06)

Discussion of proposal to transfer national forests to Interior department.

Forest policy assures greatest timber growth. (In American forestry, Jan. 1922, v. 28: 48-49)

Forest service surveys Mississippi watersheds. (In Official record of United States Department of agriculture, July 27, 1927, v. 6, no. 30: 1, 3)

Graves, H. S. National lumber and forest policy. 1919. 14 p. (United States Agriculture dept., Office of the Secretary, Circ. 134)

—— Policy of forestry for the nation. 1919. 11 p. (United States Agriculture dept., Office of the Secretary, Circ. 148)

Graves, H. S. and Clapp, E. Policy of the Forest service in regard to coöperation with private owners. (In Timberman, Dec. 1911, v. 13, no. 2: 50-53)

Hall, W. L. Moving towards a broad forestry policy; also considerations bearing upon a national forestry policy, by W. B. Greeley. Chicago, Central states forestry league, 1921. 16 p.

McNary bill [editorial]. (In American forests and forest life, Feb. 1924, v. 30: 99-100)

Criticism of the McNary bill

Nation wide study of forest taxation. (In Official record of United States Department of agriculture, Nov. 11, 1925, v. 4, no. 45: 1-2)

New forestry act. (In American forestry and forest life, July 1924, v. 30: 392, 414)

New forestry bill establishes policy. (In Official record of United States Department of agriculture, July 2, 1924, v. 3, no. 27: 2, 8)

Pinchot, Gifford. Chief forester visits Colorado. (In Southern lumberman, Mar. 20, 1909, v. 59, no. 704:29-30)

Policies of Forest service

—— Government's forest policy. (In Cyclopedia of American agriculture, ed. by L. H. Bailey, 1909. v. 4:152-54)

Secretary [Wallace] reviews President's [Harding] speech. (In Official record of United States Department of agriculture, Sept. 5, 1923, v. 2; no. 36:1)

Upholds Forest service work in Alaska.

Toumey, J. W. State's responsibility in the forest program. (In American forestry, Dec. 1921, v. 27:784-85)

Discussion of Capper and Snell bills.

U. S. Agriculture dept. Memorandum. Summary of important plans and activities of Department. 1913-1918.

Mimeographed and in library of Department.

—— —— Plans and activities of the Department of agriculture, 1913-1919. 1919. 67 p.

Mimeographed.

Forest service, p. 59-61.

—— —— Program of work of Department of agriculture, fiscal year, 1915. 1914. 278 p.

—— —— Same. 1916. 447 p.

—— —— Same. 1917. 502 p.

—— —— Same. 1919. 617 p.

—— —— Report of committee of bureau chiefs [relative to economy in research, coöperation in scientific work, etc.] 1911. 60 p.

—— —— Report of the Secretary of agriculture, 1862-date.

See Forest service in index to each volume.

—— —— Yearbook, 1894-date. 1895-date.

See index in each annual volume.

—— *Forest. service.* Field program, 1905-1922. Published 1905-July 1910, monthly, Oct. 1910-Dec. 1911, quarterly, 1911-date, semiannually

Superseded by Service directory, 1922-date.

—— —— First conference on national program of forestry. 1919. 13 leaves.

Mimeographed.

—— —— Principles and procedure governing the classification and segregation of agricultural and forest lands in the national forests. 1914. 21 p.

ORGANIZATION, PERSONNEL, AND BUSINESS PROCEDURE

Cecil, G. H. Forest service, lumber sales contracts. (In Lumberman, Sept. 1914, v. 15, no. 11 : 69-72)

Headley, Roy. Budgets and financial control in the national forest service. (In American academy of political and social sciences, May 1924, v. 113 : 51-56)

U. S. Agriculture dept. Five and 10 per cent increase; statement showing number of persons, grades or character of positions, original rates of compensation and increased rates of compensation, carried on various rolls of bureaus and offices of this Department on Sept. 30, 1917. 1917. 32 p. (65th Cong., House doc. 482) Serial 7445

Forest service, p. 14-18.

—— —— List of workers in subjects pertaining to agriculture, 1914-1922.

—— —— *Division of accounts and disbursements.* Standard classification of activities and procedure in Forest service accounting. 1912. 47 p.

—— —— *Publications division.* Organization of Department, 1901-1913, 1901-1912 published as Circular No. 1 of Division of Publications.

—— *Congress. House. Committee on agriculture.* Recommendations of Secretary of agriculture, Hearings, Mar. 2, 1922. 33 p. Serial V

Legal status of bureaus of Department of agriculture, Forest service, p. 8-9.

—— *Forest service.* Green book; instructions relating to salaries, travel and field expenses, cost keeping, property accountability, and the preparation of accounts in accordance with fiscal regulations. 1908. 47 p.

—— —— Information regarding employment on the national forests. 1905-date.

Reissued from time to time with revisions.

—— —— Manual of procedure for the Forest service in Washington and in district offices. 1908. 93 p.

—— —— National forest receipts for the benefit of schools and roads. 1913-1923.

—— —— Service directory, 1920-date.

Title differs slightly in different years.

Woolsey, T. S., Jr. Forest service revenue and organization. (In Forest quarterly, June 1916, v. 14:188-235)

PUBLICITY METHODS

Airplane photography used in making new films. (In Official record of United States Department of agriculture, Oct. 18, 1922, v. 1, no. 42:2)

Made coöperatively by Forest service.

Everard, L. C. Films from the national forests. (In Lumber, Nov. 3, 1919, v. 64, no. 18:15-16)

Motion pictures in the making. (In Official record of United States Department of agriculture, Mar. 15, 1922, v. 1, no. 11:6)

Work of Forest service featured.

Reynolds, H. A. Human equation in the forest fire problem. (In Journal of forestry, Nov. 1927, v. 25:783-801)

Educational and publicity work of the Forest service.

Simpson, F. H. Break your match in two. (In Outlook, Dec. 22, 1915, viii:971-78)

LAWS AND REGULATIONS

Kinney, J. P. Development of forest law in America. . . . New York, John Wiley and sons, 1917. 254 p.

—— Essentials of American timber law. . . . New York, John Wiley and sons, 1917. 279 p.

U. S. Agriculture dept. Office of the solicitor. Laws, decisions, and opinions applicable to the national forests. Revised and compiled by R. F. Feagans. 1916. 151 p.

—— *Forest service.* Federal and state forest laws, compiled by G. W. Woodruff, 1904. 250 p.

—— —— Law enforcement on national forests. California district. United States forest service for California, 1920. 107 p.

—— —— Revised, 1923. 102 p.
—— —— Revised, 1926. 89 p.
—— —— Trespass on national forests of Forest service district 1, by P. J. O'Brien. 1926. 88 p.
—— *Laws, statutes, etc.* Compilation of laws and regulations and decisions thereunder, relating to the creation and administration of public forest reserves. 1900. 64 p.
—— —— Compilation of laws and regulations, and decisions thereunder, relating to establishment of federal forest reserves . . . and the administration thereof. . . . 1903. 122 p.
—— —— Compilation of public timber laws, and regulations and decisions thereunder. . . . 1903. 226 p.
—— *President* (Harding). Federal forestry bill. Letter . . . regarding, proposed legislation . . . on coöperation between federal government, the states, and owners of timberlands. . . . 1923. 6 p. (67th Cong., House doc. 558)   Serial 8215

EDUCATION OF FOREST STUDENTS AND RESEARCH WORK

Aerial survey reveals resources of forests. (In Official record of United States Department of agriculture, Nov. 24, 1926, v. 5, no. 47:1, 3)

American tree association. National program of forest research, by E. H. Clapp and others. Washington, D. C. American tree assoc. 1926. 232 p.

Research agencies, p. 148-157; suggestions for an organic act for forest research in United States Department of agriculture, p. 208.

Application of new knowledge to industry. (In Chemical age, Jan. 1922, v. 30:15-16)

Forest products laboratory.

Baum, A. M. Young man, forestry and the forest service. (In Forestry Kaimin, Journal of school of forestry, University of Montana, 1922, v. 4, no. 1: 16-17, 30-31)

Burdon, E. R. Development of research work in timber and forest products. (In Royal society of arts, Journal, Mar. 7, 1913, v. 61: 438-48)

Champion, F. J. Work of the Forest products laboratory. (In New York lumber trade journal, July 15, 1926, v. 81, no. 962: 42, 106)

—— Work of the Forest products laboratory. (In Empire forester, Annual number, 1925, v. 11, no. 1: 9-14)

Cline, M. B. Wood utilization. (In Packages, June, 1908, v. 11, no. 6: 41-48)

Crissey, Forrest. Working to save wood waste. (In Saturday evening post, Feb. 3 and 17, 1912, v. 184, no. 32: 8-10; no. 34: 30)

Forest products laboratory opportunities. (In Paper, Jan. 25, 1922, v. 29: 7-9)

Forest products laboratory program and publications. (In Paper trade journal, Aug. 30, 1923, v. 77: 45-49)

Forest products laboratory work. (In Official record of United States Department of agriculture, Jan. 10, 1923, v. 2, no. 2: 5)

Forest research council. (In Science, Apr. 11, 1924, n. s. v. 59: 332-33)

Forest service offers study courses. (In Official record of United States Department of agriculture, Jan. 10, 1923, v. 2, no. 2: 5)

Frothingham, E. H. Scientific research and South Appalachian forests. (In Lumber world review, Nov. 10, 1923, v. 45, no. 9: 47-52)

Fullaway, S. V. Forest utilization a public service rendered by U. S. Forest service. (In Idaho forester, 1922, v. 4: 11-14) Forest products laboratory.

Graves, H. S. Work of the government in forest products. (In American forestry, July 1910, v. 16: 405-08)

Greeley, W. B. Demands of the national forests for technical work and trained men. (In American forestry, May 1913, v. 19: 326-30)

—— Forest products research in the United States. (In Lumber world review, Nov. 10, 1923, v. 45, no. 9: 54-56)

—— Scientific background of the forest policy of the United States. (In Science, May 23, 1924, n. s. v. 59: 449-52)

Hall, W. L. Testing laboratory for the Forest service. (In Wagon builder, Sept. 1906, v. 1, no. 6: 11-13)

Hunt, G. M. Relation of the work of the Forest products laboratory to engineering. (In Western society of engineers, Journal, May 5, 1920, v. 25: 312-29)

Jones, C. H. Forest products laboratory. (In Chemical and metallurgical engineering, Dec. 24, 1919, v. 21: 757-64)

Kellogg, R. S. Work of the Forest service which is of special value to the manufacturer. (In Manufacturer, Apr. 15, 1908, v. 20, no. 10: 295-97)

Kneipp, L. F. Technical forester in national forest administration. (In Journal of forestry, Feb. and Apr. 1918, v. 16: 155-67, 377-82)

Mathews, W. L. Forest service and you; exhibits as an aid in teaching correlated material to classes in woodwork. (In Industrial arts magazine, Aug. 1920, v. 9: 299-304)

Maudling, C. V. Research laboratory adopts modern management. (In Industrial management, Feb. 1, 1921, v. 61: 98-102)

Pinchot, Gifford. Training of a forester. 3d ed. Philadelphia, J. B. Lippincott company, 1917. 157 p.

Rue, J. D. Recent developments in forest products research in relation to forestry. (In Journal of forestry, Mar. 1926, v. 24: 237-42)

Sackett, H. S. Efficient work of the Forest service. (In Lumber world, May 15, 1910, v. 10, no. 10: 21-23)

Research work.

Secretary Wallace appoints Forest research council. (In Official record of United States Department of agriculture, Apr. 23, 1924, v. 3, no. 17: 2)

Simmons, R. E. Work of the Office of wood utilization. (In Barrel and box, Dec. 1909, v. 14, no. 10: 40D)

Skeels, Dorr. Forest ranger education. (In Proceedings of Society of American foresters, July, 1915, v. 10, no. 3: 271-83)

Society of American foresters. National program of forest research, prepared by E. H. Clapp. . . . Washington, D. C. published by American tree association for Society of American foresters, 1926. 232 p. Forest service, p. 149-157.

Entered also under American tree association.

Uncle Sam's high priced school. (In Scientific American, June 1924, v. 130: 400)

U. S. Agriculture dept. Education of forest students . . . statement of amount of money expended under the direction of the forester in the education of students. 1910. 3 p. (61st Cong., Senate doc. 443)   Serial 5658

United States Forest products laboratory. (In Automotive industry, Feb. 10, 1921, v. 44: 255)

U. S. Forest products laboratory. Forest products laboratory; a decennial record, 1910-1920 . . . Madison, Wis. Democrat printing company, 1921. 196 p.

—— *Forest Service.* Forest products laboratory, Madison, Wisconsin, Tracy and Kilgore [printers]. 1914. 21 p.

—— —— Forests products laboratory, Madison, Wisconsin. 1922. 47 p.

—— —— Forest products laboratory, supplementing exhibit . . . at Brazil centennial exposition . . . 1922-1923. 1922. 31 p.

—— —— Forestry as a profession, by E. A. Sherman. 1927. 16 p.

—— —— Profession of forestry, by H. G. Graves. 1912. 17 p.

(United States Agriculture department. Forest service circ. 207)

—— —— Review of Forest service investigations. 1913. 2 v.

Weber, A. M. District ranger training school. (In Timberman, Dec. 1926, v. 28, no. 2 : 126)

Weiss, H. F. Forest products investigations. (In Journal of forestry, July 1925, v. 23 : 565-73, 574-82)

Wells, S. D. Forest products laboratory an aid to the paper industry. (In Paper mill Aug. 7-14, 1915, v. 38, no. 32 : 1, 11; no. 33 : 11, 34, 36)

Work of the Forest products laboratory. (In Aerial age; Dec. 26, 1921, v. 14 : 374)

Years' work in Forest service industrial investigations. (In St. Louis Lumberman, May 15, 1916, v. 57, no. 10 : 12)

Zon, Raphael. Advisory committee to the Lake states forest experiment station. (In Science, May 2, 1924, n. s. v. 59 : 393)

—— Lake states experiment station and its field. (In Lumber world review, Nov. 10, 1923, v. 45, no. 9 : 61-63)

—— Lake states forest station. (In Paper, Nov. 15, 1923, v. 33 : 11-13)

FOREST EXPERIMENT STATIONS

Conducts experiments on trees [Fremont forest experiment station]. (In Official record of United States Department of agriculture, July 9, 1924, v. 3, no. 28 : 6)

Dana, S. T. Northeastern forest experiment station. (In Paper mill and wood pulp news, Apr. 12, 1924, v. 48 : 78)

—— What the Northeastern forest experiment station should aim at. (In Journal of forestry, Jan. 1923, v. 21 : 40-43)

Forbes, R. D. What the Southern forest experiment station is doing. (In Lumber world review, Nov. 10, 1922, v. 43, no. 9 : 51-55)

Forest experiment stations. (In Science, Dec. 16, 1921, n. s. v. 54 : 599)

Forest experiment stations provided. (In Official record of United States Department of agriculture, Sept. 12, 1923, v. 2, no. 37 : 1)

Forest research in the South [experiment stations]. (In Official record of United States Department of agriculture, Apr. 29, 1925, v. 4, no. 17 : 2)

Frothingham, E. H. Forest experiment station for the South. (In American forestry, Sept. 1921, v. 27 : 598-99)

Hofman, J. V. Wind River forest experiment station. (In West Coast lumberman, May 1923, v. 44 : 118-20)

Lake states forest experiment station. (In Science, Sept. 14, 1923, n. s. v. 58 : 200)

Munns, E. N. Organization and development of Federal forest experiment stations. (In Association of land grant colleges, Proceedings, 1926 : 186-95)

New forest station in Ohio. (In Official record of United States Department of Agriculture, July 13, 1927, v. 6, no. 28 : 2, 8)

Southern forest experiment station. (In Science, Nov. 18, 1921, n . s. v. 54 : 487)

—— *Congress. House. Committee on agriculture.* Forest experiment station in Pennsylvania, report to accompany S. 2516. . . . 1926. 5 p. (69th Cong., House rept. 1422)
                                                        Serial 8534

—— —— —— —— Forest experiment station, hearings on H. R. 397 and S. 2516, June 3 and 8, 1926. 1926. 25 p.
                                                        (Serial P)

—— —— —— —— Forest experiment stations in California, report to accompany S. 4156 (to authorize establishment and maintenance of forest experiment station in California and surrounding states) 1925. 4 p. (68th Cong., House rept. 1581)   Serial 8391

—— —— —— —— To authorize establishment and maintenance of forest experiment station in Ohio and Mississippi Valleys, report to accompany S. 3405. . . . 1926. 3 p. (69th Cong., House rept. 1430)   Serial 8534

—— —— *Senate. Committee on agriculture and forestry.* Establishment of forest experiment station in Pennsylvania, hearing . . . on S. 2516 for establishment and maintenance of forest experiment station in Pennsylvania and neighboring states. 1926. 8 p.

—— —— —— —— Establishment of forest experiment station in Pennsylvania, report to accompany S. 2516. . . . 1926. 4 p. (69th Cong., Senate rept. 619)   Serial 8525

—— —— —— —— Forest experiment station in Ohio and Mississippi valleys, report to accompany S. 3405. . . . 1926. 2 p. (69th Cong., Senate rept. 891)   Serial 8526

—— *Forest service.* Lake states forest experiment station. 1927. 34 p.

—— —— Northeastern forest experiment station, Amherst, Mass. Forest investigation under way in New England and New York, Apr. 1, 1924. 1924. 83 numbered leaves.

Mimeographed.

—— —— Program of work of the forest experiment stations. 1923, 1925, 1928.

Mimeographed.

—— —— Rocky Mountain forest experiment station. . . . 1926. 32 p.

FOREST RESERVATIONS, GENERAL

Ammons, E. M. Forest reservations. . . . 1910. 18 p. (61st Cong., Senate doc. 650)   Serial 5660

Ashe, W. W. Creation of the Eastern national forests. (In American forestry, Sept. 1922, v. 28: 521-25)

Baum, A. M. Why our forests are burning up. (In Outlook, Aug. 10-24, 1927, v. 146: 472-74, 508-10, 540-43)

Attack on Forest service.

Beaman, D. C. Abuses of the forest reserve system as at present administered. Denver, Colorado, no publisher, 1907. 16 p.

Bishop, L. L. Allegheny national forest. (In Forest leaves, June 1922, v. 18: 139-41)

Boerker, R. H. D. Our national forests, a short popular account of the work of the United States Forest service on the national forests. New York, Macmillan company, 1918. 238 p.

Brown, Ralph. Acquisition of land in the White mountains, under the Weeks law. (In Empire forester, 1916, v. 2: 11-17)

Buck, Shirley. Uncle Sam's pure water bureau; forest rangers protect from contamination streams that furnish cities with water. (In Suburban life, Nov. 1913, v. 17:255)

Colonel Greeley rejects Baum charges. (In American forestry, Nov. 1927, v. 33:647-48, 702-03)

Cornwall, G. M. Utilizing troops in the national forests. (In American forestry, Oct. 1911, v. 17:587-89)

Eastern forests reserve enlarged. (In Official record of United States Department of agriculture, June 14, 1922, v. 1, no. 24:3)

Eight national forests created [on military reservations]. (In Official record of United States Department of agriculture, Apr. 22, 1925, v. 4, no. 16:5)

Extension of forest lands is discussed. (In Official record of United States Department of agriculture, Apr. 2, 1924, v. 3, no. 14:1, 5)

Forest service acquires new land. (In Official record of United States Department of agriculture, Sept. 19, 1923, v. 2, no. 38:3)

Describes new methods used by Forest service.

Forest service contacts. (In Official record of United States Department of agriculture, May 23, 1923, v. 2, no. 21:2)

Forests and forests. (In Outlook, June 1924, v. 140:284-85)

On the administration and use of the national forests.

Forests for the nation. (In American forestry, Feb. 1924, v. 30:93)

Additional legislation supplementary to Weeks law urged.

Leavitt, Clyde. Methods in determining reserve boundaries. (In Forest quarterly, Dec. 1906, v. 4:274-81)

National forests, June 30, 1926. (In United States Agriculture dept., Yearbook, 1926:794-95)

Norcross, T. W. Where your treasure lies. (In Outdoor America, March, 1925, v. 3, no. 8:46-47, 49)

Forest reservations.

Plan for eastern national forests gains strength. (In American forests and forest life, Feb. 1925, v. 31:105-06)

Prentiss, A. M. Trees where soldiers trained. (In Outlook, Dec. 2, 1925, v. 141:518-19)

16

Rath, Filibert. Administration of forest reserves. (In Forestry and irrigation, 1902, v. 8 : 191-93, 241-44, 279-82)

Report on national park and forest boundaries. (In American forests and forest life, Dec. 1925, v. 31: 768)
With maps.

Richards, Edward. National forests and the Northern Pacific. (In Nation, July 16, 1924, v. 119 : 71-73)

U. S. Agriculture dept. National forests of California, by R. W. Ayres and W. Hutchinson, 1927. 34 p. (United States Agriculture dept. Misc. circ. 94)

—— —— National forests of New Mexico. 1922. 21 p. (U. S. Agriculture dept. circ. 240)

—— —— National forests of Wyoming. 1927. 26 p. (United States Agriculture dept. Misc. circ. 82)

—— —— Report of national forest funds, report of contributions on account of coöperative work with the Forest service . . . fiscal year 1918. . . . 1918. 14 p. (65th Cong., House doc. 1549) Serial 7584

—— *Congress. House. Committee on agriculture.* Appalachian forest reserve. . . . Report (to accompany H. R. 19573). 1906. 21 p. (59th Cong., 1st sess. House rept. 4399) Serial 4908

—— —— —— —— Authorizing Secretary of agriculture to coöperate with territories in supervision and extension of forests, report to accompany H. J. Res. 52. . . . 1926. 3 p. 69th Cong., House rept. 434) Serial 8532

—— —— —— —— Creation of game refuges in Ozark national forest, Ark., report to accompany H. R. 12192. . . . 1925. 2 p. (68th Cong., House rept. 1438) Serial 8391

—— —— —— —— Exchange of forest lands. Hearings on H. R. 10679, H. R. 10680, H. R. 10681, Jan. 6, 22, 1925. 1925. 9 p. Serial Y 2 pts

—— —— —— —— Hearings . . . on bills having for their object the acquisition of forest and other lands. . . . 1909. 143 p.

—— —— —— —— Purchase of forest lands. Hearings, Jan. 13-14, 1922. 1922. 58 p. Serial N

—— —— —— *Committee on expenditures in Department of agriculture.* Report . . . 1907. 64 p. (59th Cong., 2d sess. House rept. 8147) Serial 5065
Forest reserves, p. 13-19.

—— —— —— *Committee on judiciary.* . . . Power of federal government to acquire lands for national forest purposes. . . . 1908. 42 p. (60th Cong., House rept. 1514)
Serial 5226

—— —— —— *Committee on public lands.* Forest reserve administration . . . report on . . . bill to transfer certain forest reserves to the control of Department of agriculture. . . . 1902. 15 p. (57th Cong., 1st sess. House rept. 968)
Serial 4402

—— —— —— —— Same. Views of minority. 7 p.

—— —— —— —— General forest exchange bill; hearings . . . on H. R. 9539, a bill for the consolidation of forest lands within the National forests. 1920. 137 p.

—— —— —— —— Public forest reservations. . . . 1893. 5 p. (53d Cong., 1st sess. House rept. 78)   Serial 3157

—— —— —— —— Public forest reservations. . . . 1894. 23 p. (53d Cong., 2d sess. House rept. 897)   Serial 3271

—— —— —— —— Public forest reservations. . . . 1896. 20 p. (54th Cong., 1st sess. House rept. 1593)   Serial 3462

—— —— *Senate.* New forest reservations . . . letters and memorials relating to new forest reservations. 1897. 11 p. (55th Cong., 1st sess. Senate doc. 68)   Serial 3562

—— —— —— *Committee on agriculture and forestry.* Report [on public forest reservations]. 1892. 12 p. (52d Cong., 1st sess. Senate rept. 1002)   Serial 2915

—— —— —— *Committee on forest reservations and the protection of game.* Acquiring national forests in Southern Appalachian and White Mountains. Report (to accompany S. 4825). 1908. 15 p. (60th Cong., Senate rept. 459)
Serial 5219

—— —— —— *Committee on public lands and surveys.* Boundaries of Yellowstone Park, report pursuant to S. Res. 237. 1927. 3 p. (69th Cong., Senate rept. 1714)
Serial 8687

—— —— —— —— National forests and public domain, hearings pursuant to S. Res. 347. 1926. 16 pts. 3 vols.

—— *Forest service.* Bryce Canyon national monument, Utah, Powell national forest. . . . 1927. 8 p.

—— —— Country's forests. 1914. 14 p.

—— —— Handbook for campers in the national forests of California. 1915. 148 p.

—— —— Same. 1920. 48 p.

—— —— Same. 1921. 48 p.

(United States Agriculture dept. circ. 185.)
Contains much information on the work of the Forest service.

—— —— Ideal vacation land—the national forests of Oregon. 1924. 56 p.
—— —— Importance of forestry and the national forests. 1924. 16 p.
—— —— Same. 1925. 16 p. (United States Agriculture dept. Misc. circ. 15)
—— —— Instructions. 1905-date.

Issued irregularly, from Washington office. Local instructions are issued also by district foresters.

—— —— Inter-mountain district forest statistics. 1925. 64 p.
—— —— Mountain vacation land—the Washington national forest. 1920. 10 p. (United States Agriculture dept. circ. 132)
—— —— Same. Revised. 1923. 15 p.
—— —— National forest areas. 1905-date.
—— —— National forest manual. 1911-date.

Contains regulations and instructions, different subjects covered in each edition.

—— —— National forest resources of Utah. 1926. 27 p. (United States Agriculture dept. Misc. circ. 71)
—— —— National forests of Arizona; prepared by Southwestern district of Forest service. 1924. 19 p. (United States Department of agriculture, circ. 318)
—— —— National forests of Arkansas. 1909. 5 p.
—— —— National forests of Idaho; prepared by Intermountain and northern districts of Forest service. 1926. 34 p. (United States Agriculture dept. Misc. circ. 61)
—— —— National forests of the Southern Appalachians. 1923. 22 p.
—— —— National forests of Wyoming. . . . 1927. 26 p. (United States Agriculture dept. Misc. circ. 82)
—— —— Northern Pacific land grant. Report of the United States Department of agriculture Forest service to . . . Northern Pacific railway company. 1924. 327-356 p. Pt. 5

of Hearings on Northern Pacific land grant before House Committee on public lands.

—— —— Olympic national forest: Its resources and their management, by Findley Burns. 1911. 20 p. (United States Forest service bulletin 89)

—— —— Oregon caves. . . . Siskiyou national forest. 1926. 16 p.

—— —— Plan for the development of the village of Grand Canyon, Arizona, by F. A. Waugh. 1918. 23 p.

—— —— Purchase of land for national forests under act of Mar. 1, 1911. Weeks law. 1924. 15 p. (United States Agriculture dept. circ. 313)

—— —— Purchase of land under the Weeks law in the Southern Appalachian and White mountains. 1911-1924.

—— —— Use of national forest reserves. 1905. 142 p.

—— —— Use of national forests. 1907. 42 p.

—— —— What national forests mean to intermountain region, by F. S. Baker. 1925. 21 p. (United States Agriculture dept. Misc. circ. 47)

—— —— Wichita national forest and game preserve, by S. M. Shanklin and J. E. Scott. 1925. 11 p. (United States Agriculture dept. Misc. circ. 36)

—— *General land office.* Regulations governing forest reserves established under section 24 of act Mar. 3, 1891. 1897. 14 p.

—— —— Same. 1898. 16 p.

—— —— Same. 1900. 17 p.

—— *Geological survey.* Survey of forest reserves. Letter . . . relative to survey of public lands that have been designated as forest reserves by executive proclamation. 1897. 16 p. (55th Cong., 2d sess. Senate doc. 47). Serial 3592

—— *Interior dept.* Establishment of forest reservations . . . letter from Secretary of interior transmitting . . . copies of orders of President of Feb. 22, 1897, and copies of papers relating to establishment of forest reservations. 1897. 36 p. (55th Cong., 1st sess. Senate doc. 21) Serial 3590

—— *National forest reservation commission.* Hayden national forest, Colorado, . . . proposed addition of public lands to Hayden national forest, Colorado. 1926. 4 p. (69th Cong., House doc. 293) Serial 8579

—— —— Payette national forest, Idaho . . . proposed addition of public lands to Payette national forest, Idaho. 1926. 4 p. (69th Cong., House doc. 294) Serial 8579

—— —— Program of purchase of eastern national forests under the Weeks law. 1920. 23 p.

—— —— Report, 1911-date.

—— —— Resolutions adopted Jan. 8, 1927, proposing addition of public lands to Challis, Idaho and Sawtooth national forests in Idaho, Missoula and Helena national forests in Montana, Colville national forest in Washington, and Wyoming national forest in Wyoming. . . . 1927. 11 p. (69th Cong., House doc. 668)   Serial 8735

Van Name, W. G. Danger to Crater Lake national park. (In Science, July 23, 1926, n. s. v. 64: 91)

Forest Management

Administrative control of forests and disposition of national timber. (In Timberman, Mar. 1913, v. 14, no. 5: 31-32)

Ammons, E. M. Forest reservations. 1910. 18 p. (61st Cong., Senate doc. 650)   Serial 5660

A criticism of the system of control over forest reserves, and of their administration.

Anderson, Mark. Politics and science as affecting public land management. (In Journal of forestry, Nov. 1927, v. 25: 889-92)

Barnes, W. C. Biter bitten. (In American forests and forest life, Mar. 1924, v. 30:144-45, 192)

Brill, E. C. Uncle Sam, dealer in Christmas trees. (In St. Nicholas, Dec. 1923, v. 51:206-07)

Bruff, J. R. Methods of reconnaissance on the national forests. (In Nebraska' university, Forest club annals, 1912, v. 4: 74-80)

Brundage, M. R. and Berry, J. R. Estimating the cut on small sales of government timber. (In Journal of forestry, May 1923, v. 21: 483-91)

Cleveland, Treadwell. National forest work and the South, what the government is doing to promote conservative lumbering. (In Southern lumberman, Dec. 25, 1905, v. 51, no. 603: 52-54)

Dyar, W. W. Forest service and mining in the national forests. Denver, Colorado, no publisher, 1909. 15 p.

Eddy, J. A. Forest service operating the forest reserves, as it affects welfare of the people. Denver, Colorado, no publisher, 1909. 28 p.

Attack on administration of Forest service.

Enderslee, W. J. Land classification in Arizona. (In Empire forester, 1916, v. 2 : 55-58)

Freeing the forest reserves from predatory animals. (In Scientific American, Dec. 1918, v. 7, no. 6 : 571-72)

Greeley, W. B. Forest management on federal lands. (In Journal of forestry, Mar. 1925, v. 23 : 223-35)

—— Forest service is sticking to its job. (In Outlook, Mar. 4, 1925, v. 139 : 336-39)

Guthrie, J. W. Handling Uncle Sam's forest properties. (In West Coast lumberman, May 1, 1924, v. 45 : 61, 64)

How Uncle Sam's woodlot helps pay for its keep; a story in pictures. (In Nation's business, Nov. 1916, v. 4, no. 11 : 24-25)

Kinney, J. P. Forestry administration on Indian reservations. (In Journal of forestry, Dec. 1921, v. 19 : 836-43)

Largest pulp timber sale by department [Forest service]. (In Official record of United States Department of agriculture, Aug. 29, 1923, v. 2, no. 35 : 1, 5)

Moles, H. S. Ranger district no. 5. Boston, Mass. Spencerian press, 1923. 350 p.

Munger, T. T. Logging national forest timber in Douglas fir region. (In Lumber world review, June 10, 1922, v. 42, no. 11 : 28-30)

Orr, Raymond. Timber sales administration. (In Gopher Peavey forestry club, University of Minnesota, 1922, v. 2 : 16-17)

Redington, P. G. Relation of the Forest service to the mining industry. (In Mining world, Apr. 24, 1909, v. 30, no. 17 : 789-90)

Riley, S. Forest service and the prospector. (In Engineering and mining journal, July 26, 1913, v. 96 : 175-76)

Smith, H. A. How the public forests are handled. (In U. S. Agriculture dept., Yearbook, 1920 : 309-30)

Tillotson, C. R. Nursery practice on the national forests. 1917. 86 p. (U. S. Agriculture dept., Bulletin 479)

Timber sale policies upheld by Secretary [of agriculture]. (In Official record of United States Department of agriculture, Apr. 13, 1927, v. 6, no. 15 : 3, 6)

Timber users turn to national forests. (In Official record of United States Department of agriculture, Jan. 5, 1927, v. 6, no. 1 : 3, 8)

Handling of timber sales by Forest service.

U. S. Agriculture dept. . . . Expenditures for national forest administration. . . . 1912. 49 p. (62d Cong., House doc. 681)   Serial 6326

—— —— Government forest work in Utah. 1921. 31 p. (United States Agriculture dept. Circ. 198)

—— —— Our national elk herds, by H. S. Graves and E. W. Nelson. 1919. 34 p. (U. S. Agriculture dept. Circ. 51)

—— *Congress. Senate. Committee on agriculture and forestry.* Acquisition of dams in Minnesota national forests. Hearing . . . H. R. 292. Apr. 24, 1926. 1926. 3 p.

—— —— —— —— Dams in Minnesota national forest. Report to accompany H. R. 292. . . . 1926. 3 p. (69th Cong., Senate rept. 655)   Serial 8525

—— —— —— —— Game sanctuaries in national forests, report to accompany S. 1147. . . . 1926. 3 p. (69th Cong., Senate rept. 886)   Serial 8526

—— —— —— *Committee on public lands and surveys.* Herrick timber contract; Malheur national forest, Oregon. Hearings . . . on S. Res. 332, to investigate all matters relating to contract between Fred Herrick and Forest service for purchase of timber. Jan. 28, and Feb. 24, 1927. 1927. 894 p.

A private case, but the hearings contain much information on administration of Forest service.

—— *Forest service.* First aid manual for field parties, by H. W. Barker. 1917. 98 p.

—— —— Homestead in the national forests. 1917. 12 p.

—— —— National forest manual. 1911-date.

Now in loose leaf form, with different editions, sometimes more than one a year.

—— —— Pisgah national game preserve regulations. 1917. 14 p.

—— —— Same. 1917. 11 p.

—— —— Same. 1921. 11 p. (United States Agriculture dept. Circ. 161)

—— —— Principles and procedure governing the classification and segregation of agricultural and forest lands in the national forests. 1914. 23 p.

—— —— Rules and regulations permitting prospecting, development and utilization of the mineral resources of lands acquired under act of Mar. 1, 1911. 1917. 19 p.

—— —— Use book. Manual of information about the national forests, 1906-date.

Published irregularly and with special editions for special subjects, such as grazing and water power.

—— —— Water power projects, telephone, telegraph, power transmission lines on the national forests. Regulations. . . . 1915. 90 p.

Wolff, M. H. Land classification as part of the national forest work in Montana and north Idaho. (In Forestry Kaimin, Journal of school of forestry, Montana state university, 1925: 65-68, 95-110)

Woodbury, T. D. Logging national forest timber in California. (In Lumber world review, July 25, 1922, v. 43, no. 2: 27-29)

—— Uncle Sam—lumberman. (In West Coast lumberman, May 1, 1925, v. 48, no. 566: 110-12)

Yarnall, I. T. White Mountain national forest. (In N. H. forests, Sept. 1927, v. 4, no. 3: 1-2)

Grazing Rights and Usages

Adjustment of grazing fees on Nevada forests. (In National wool grower, July 1927, v. 17: 12-13)

Association adopts principles of grazing legislation. (In American forests and forest life, May 1926, v. 32: 279-81)

Authier, G. F. Both sides of the range controversy. (In American forests and forest life, Dec. 1925, v. 31: 716-17)

Barnes, W. C. Carrying capacity of range increased by systematic use. (In American Hereford journal, May 1, 1922, v. 13: 24-27)

—— Forest service and the stockman in 1917. (In National wool grower, Feb. 1918, v. 8, no. 2: 15-18)

—— Same. (In American sheepbreeder, Feb. 1918, v. 38, no. 2: 88-89)

—— Sheepmen on the national forests. (In National wool grower, Feb. 1921, v. 11, no. 2: 21-22, 33-35)

—— Stockmen and forest ranges. (In Breeder's gazette, Mar. 22, 1923, v. 80: 394-95)

—— Uncle Sam range cowman. (In Breeder's gazette, Aug. 16, 1917, v. 72: 203-04)

Betts, F. E. Sheep in the national forests. (In Breeder's gazette, Dec. 31, 1925, v. 88, no. 27: 856-57; Jan. 7, 1926, v. 89, no. 1: 15-16)

Butler, O. M. Shall the stockmen control the national forests? (In American forests and forest life, Sept. 1925, v. 31: 519-22, 574)

Cabinet officers disapprove Stanfield grazing bill. (In American forests and forest life, March 1926, v. 32: 149-50)

Chapline, W. R. Goats on national forests. (In Angora and milk goat journal, Jan. 1919, v. 8, no. 5: 9-10)

Chapman, H. H. Grazing menace on our national forests. (In American forests and forest life, Feb. 1926, v. 32: 85-88)

Cobbs, J. L. Bringing in the breeds. (In Outing magazine, Jan.-Feb. 1919, v. 73: 177-80, 250-54)

Col. Greeley on forest problems. (In Producer, Oct. 1925, v. 7: 18-19)

Dillon, Richard. Comes now the plaintiff. (In Producer, Apr. 1924, v. 5: 5-9)

Criticism of Forest service policy toward grazing charges and privileges.

Forest forage fees, report by D. D. Casement. (In Official record of United States Department of agriculture, Nov. 24, 1926, v. 5, no. 47: 6)

Forest forage resources to come under new rules. (In Official record of United States Department of agriculture, Dec. 19, 1923, v. 2, no. 51: 3)

Gill, Tom. Stanfield grazing bill is dehorned. (In American forests and forest life, Apr. 1926, v. 32: 203-04)

Grazer and the government. (In Outlook, Apr. 14, 1926, v. 142: 556-57)

Grazing-fee question settled by Secretary. (In Official record of United States Department of agriculture, Feb. 16, 1927, v. 6, no. 7: 1, 6)

Grazing fees, 1926, part will be waived. (In Official record of United States Department of agriculture, Jan. 6, 1926, v. 5, no. 1: 1-2)

Grazing legislation. (In Producer, Apr. 1926, v. 7: 17-18)

Grazing on the national forests. (In Producer, Dec. 1925, v. 7: 19-20)

Grazing versus forestry. (In Journal of forestry, Apr. 1926, v. 24: 378-411)

Greeley, W. B. Grazing administration and charges on the national forests. (In National wool grower, Feb. 1924, v. 14: 30-31, 44-45)

—— Issue between grazing and forestry. (In World's work, Aug. 1926, v. 52 : 447-53)

—— Stockmen and the national forests. (In Saturday evening post, Nov. 14, 1925, v. 198 : 10-11+)

Hagenbarth, F. J. National forest radicals. (In National wool grower, Dec. 1925, v. 15 : 13-14)

How national forest administration benefits water users. (In Reclamation record, June 1916, v. 7, no. 6 : 269-70)

Methods of grazing management.

How the Forest service is developing the range. (In National wool growers, Feb. 1915, v. 5, no. 2 : 9-11, 19)

Investigation of Uncle Sam's ranges. (In American forests and forest life, Oct. 1925, v. 31 : 611-12)

Jardine, J. T. Range improvement and improved methods of handling stock in national forests. (In Society of American foresters, Proceedings, 1912, v. 7 : 160-67)

Korstian, C. F. Grazing practice on the national forests. (In Scientific monthly, Sept. 1921, v. 13 : 275-81)

National forests in danger. (In Agricultural review, Mar. 1926, v. 19 : 16-17)

National forests or stockmen's profits? (In Nation, Sept. 23, 1925, v. 121 : 321-22)

New grazing regulations on national forests. (In Official record of United States Department of agriculture, Feb. 24, 1926, v. 5, no. 8 : 2)

Parks, H. L. National forests grazing fees. (In Breeder's gazette, Jan. 28, 1926, v. 89 : 98, 99, 101)

Peters, T. M. Stockman and the federal land. (In Outlook, Apr. 14, 1926, v. 142 : 568-69)

Potter, A. F. Administration of grazing in national forests. . . . 1913. Denver, American livestock association, 1913. 15 p.

—— Forest service side of grazing problems. (In National wool grower, May 1920, v. 10, no. 5 : 14-16)

—— How the Forest service has helped the livestock man. (In American national livestock association, Proceedings, 1915 ; 48-55)

—— How the Forest service has helped the stockman. (In American forestry, Mar. 1918, v. 24, no. 291 : 165-69)

Potter, E. L. Stockmen and the forest service. (In Breeder's gazette, Mar. 18, 1926, v. 89 : 330)

—— Takes issue with Colonel Greeley. (In Producer, Dec. 1925, v. 7: 17-18)

Review of Forest service grazing, by D. D. Casement. (In Official record of United States Department of agriculture, Aug. 26, 1925, v. 6, no. 34:2)

Revised grazing bill. (In National wool grower, Apr. 1926, v. 16: 14-16)

Senate 2584 as amended.

Rockwell, R. F. Grazing cattle on the National forest reserve. (In Overland monthly, Feb. 1926, n. s. v. 84: 35-36)

—— Rachford refuses to lower forest fees. (In Producer, Oct. 1927, v. 9: 14-15)

Sampson, A. W. Grazing, recreation and game in the forests. (In California countryman, Mar. 1926, v. 12: 5, 27-28)

Secretary Jardine's statement upon forest grazing administration. (In National wool grower, Dec. 1925, v. 15: 25-27)

Senate committee hearings on Stanfield grazing bill. (In National wool grower, Mar. 1926, v. 16: 13-14)

Smith, G. A. Attack on the forest service grazing policy. (In Journal of forestry, Feb. 1926, v. 24: 136-40)

Stockmen and the national forests. (In American forests and forest life, Sept. 1925, v. 31: 548)

Stockmen pass resolution to muzzle research. (In American forests and forest life, Mar. 1926, v. 32: 167)

Stockmen submit public land recommendations. (In Producer, Sept. 1925, v. 7: 7-8)

Stockmen would quarter the national forests. (In American forests and forest life, Nov. 1925, v. 31: 667-68)

Storm, E. N. Salting on the forest range. (In Producer, Feb. 1920, v. 1, no. 9: 12)

To save the forests from the stockmen. (In Producer, Nov. 1925, v. 7: 16-17)

U. S. Congress. House. Committee on agriculture. Authorizing and directing Secretary of agriculture to waive one-half of grazing fees for use of national forests. Report to accompany H. J. Res. 375, . . . 1925. 2 p. (68th Cong., House rept. 1617)   Serial 8391

—— —— —— —— Grazing fees. . . . Hearings, Thursday, April 1, 1920. 1920. 30 p.

Grazing fees, p. 1-12.

—— *Congress. Senate. Committee on agriculture and forestry.* Forest experiment stations. Hearings . . . on S. 824, a bill to establish and maintain a forest experiment station in the State of Florida and on S. 1667 to authorize the purchase of lands in Florida for an experimental and demonstration forest for production of naval stores. 1924. 8 p.

—— —— *Senate. Committee on public lands and surveys.* Grazing facilities on public lands: hearings on S. 2584, Feb. 15–Mar. 11, 1926. 1926. 632 p.

—— *Forest service.* Handling of sheep on the national forests. 1920. 21 p.

—— —— Range management on national forest, by J. T. Jardine and Mark Anderson. 1919. 98 p. (United States Agriculture dept. Bulletin 790)

—— —— Revised regulations and instructions in reference to grazing. 1905. 16 p.

—— —— Saving livestock from starvation on southwestern ranges, by C. L. Forsling. 1924. 22 p. (United States Department of agriculture, Farmers' bulletin 1428)

—— —— Story of the range, by W. C. Barnes. 1926. 60 p.

Reprinted from Pt. 6 of Hearings before Senate committee on public lands and surveys, 1926.

—— —— United States Forest service experiment station news, Dec. 12, 1923-date.

Mimeographed weekly publication. Discusses work of the several forest experiment stations.

FOREST FIRE PREVENTION AND CONTROL

Airplane forest patrol makes good western record in 1921. (In Engineering news, Jan. 26, 1922, v. 88: 163)

Barnes, W. C. Girl behind the fire line. (In American forestry, Jan. 1923, v. 29: 29-32)

Work in a local forest supervisor's office.

—— Our national bonfires. Ten million acres of forest lands are burned over annually in the United States for lack of state coöperation with the federal Forest service. (In Harper's weekly, Aug. 7, 1909, v. 53, no. 2746: 8-9)

—— Trailing a fire bug. (In American forestry, May 1923, v. 29: 265-69)

Boyce, C. W. Aerial forest fire patrol in Oregon and California. (In Journal of forestry, Nov. 1921, v. 19: 771-75)

—— Aerial smoke chase. (In Timberman, June 1922, v. 23: 156-58)

Brereton, C. V. Tell it to the judge. (In Overland monthly, July 1923, n. s. v. 81:19-21, 44, 47)

Burns, Findley. Safety first in the national forests. (In Red cross magazine, Jan. 1917, v. 12, no. 1: 33-34)

Chestnut, M. V. Radio telegraphy is great aid in fighting forest fires. (In Wireless age, Aug. 1922, v. 9: 62-64)

Clapp, E. H. Fire protection in the national forests. (In American forestry Oct.-Nov. 1911, v. 17: 573-84, 652-57)

Clepper, H. E. Lookout on the hill. (In American forestry, Apr. 1924, v. 30: 204-06)

Cummings, Ralph. The fire fighters; a story of the Forest service. (In Sunset magazine, Oct.-Dec. 1921, v. 47, no. 4: 17-19, 80-89; no. 5: 38-40, 82-90; no. 6:42)

Dacy, G. H. Forest detectives. (In Popular mechanics, Aug. 1921, v. 36, no. 2: 211-13)

—— Sherlock Holmes of the forests. (In American forestry, Feb. 1922, v. 28: 72-75)

Dahl, H. L. How forest rangers guard against fires. (In California cultivator, May 5, 1923, v. 60: 519, 527)

—— How forest rangers protect Uncle Sam's forests. (In Overland monthly, April 1910, n. s. v. 55:357-61)

—— Protecting our timber resources; using the heliograph to fight forest fires. (In Scientific American supplement, July 21, 1917, v. 84:36)

Deering, R. L. Airplane forest fire patrol in California, 1920. (In Gopher Peavey forestry club, University of Minnesota, 1922, v. 2: 21-23)

Discontinuing the airplane forest patrol. (In Journal of electricity, Jan. 15, 1922, v. 48: 45)

Duggs, L. L. Fighting forest fires from the air. (In Outlook, Jan. 26, 1921, v. 127: 138-42)

Edholm, C. L. Trapping a forest fire bug. (In Technical world magazine, June 1911, v. 15: 382-89)

Fighting seas of fire in timber lands. (In Popular mechanics, Jan. 1920, v. 41: 18-20)

Forest patrols of the air. (In World's work, Dec. 1919, v. 39: 177-84)

Forest service fire prevention plans. (In Timberman, June 1922, v. 23: 142)

Frost, Stanley. Forestry from the air. (In American forestry, May 1921, v. 27: 278-80)

Goldsmith, L. C. Air patrol. (In Sunset, Sept. 1927, v. 59: 40-41)

Graves, H. S. Torch in the timber, it may save the timberman's property but it destroys the forests of the future. (In Sunset, Apr. 1920, v. 44, no. 4: 37-40)

Greeley, W. B. Fires on the national forest. (In American forestry, Jan. 1922, v. 28: 49-50)

Guthrie, J. D. Then and now. (In American forestry, Jan. 1923, v. 29: 51-52)

Forest fire patrol.

Harry, D. Fire eagles: how army planes and pilots keep check on forest fires. (In Sunset, Sept. 1920, v. 45: 104-05)

Hobson, D. G. How the timber sleuths track fire bugs. (In Illustrated world, Sept. 1922, v. 38: 75-76)

Hough, Donald. Canoe fire department. (In Outing, Dec. 1921, v. 79: 107-09)

Hutchinson, Wallace. Eyes of the forest. (In American forestry, Aug. 1922, v. 28: 461-68)

—— Forest fires, a national problem. (In American forestry, Nov. 1921, v. 27: 675-83)

Jamison, J. H. Two square inches of hide delivered the fire bug into the hands of the woods detective. (In Outing, Feb. 1923, v. 81: 212-14)

Keeping forest fires at bay. (In American industries, Mar. 1924, v. 24: 30)

Lane, D. R. Fire fighting with tractors. (In Sunset, Oct. 1927, v. 59: 27+)

Larsen, J. A. Smoky trail. (In American forestry, July 1923, v. 29: 394-97)

Laut, A. C. Fire protection of United States Forest service. (In American forestry, Nov. 1913, v. 19: 711-20)

Leveaux, C. M. Forest guard work on the national forests. (In Michigan agricultural college, Forestry club, annals, 1917, v. 2: 25-28)

Marsh, S. H. Minute men in fire protection. (In American forests and forest life, Sept. 1926, v. 52: 524-26)

Mitchell, G. E. Prevention of forest fires. (In American review of reviews, July 1911, v. 44: 64-68)

Operation of aeroplanes over forested areas. (In Aerial age, Mar. 13, 1922, v. 15: 11)

Operations of the aerial forest patrol, 1922. (In Aviation, Feb. 26, 1923, v. 14: 241-44)

Peck, A. S. Protecting the national forests from fire. (In Safety engineering, Apr. 1915, v. 29, no. 4: 293-98)

Predict fires in Montana forests. (In Official record of United States Department of agriculture, Dec. 26, 1923, v. 2, no. 52: 4)

Prevention cheaper than fighting fire. (In Official record of United States Department of agriculture, Aug. 10, 1927, v. 6, no. 32: 3)

Coöperative work on Cape Cod.

Protecting our wealth in timber. (In Radio broadcast, Nov. 1923, v. 4: 9-10)

Pulaski, E. C. Surrounded by forest fires. (In American forestry, Aug. 1923, v. 29: 485-86)

Radio talks on prevention of forest fires broadcast by Forest service. (In Official record of United States Department of agriculture, July 25, 1923, v. 2, no. 30: 2)

Redington, P. G. Aerial fire patrol on the national forests. (In Aviation, Aug. 31, 1925, v. 19: 248-49)

—— Use of the airplane in forest protection. (In California citrograph, Jan. 1922, v. 7: 67, 100)

Spotting forest fires before they happen. (In Official record of United States Department of agriculture, June 25, 1924, v. 3, no. 26: 2)

Sterling, E. A. Forest fires and the railways. (In Engineering magazine, Apr. 1912, v. 43: 111-14)

Stoddard, L. E. Airplane patrol. (In St. Nicholas, July 1921, v. 48: 771-75)

Tillotson, C. E. Federal coöperation with the States in protection against forest fires. (In N. Y. State college of forestry, Forest protection conference, 1926: 52-57)

Townsend, A. P. Aerial forest fire patrol. (In Forestry Kaimin, Journal of school of forestry, University of Montana, v. 4, no. 1: 15, 36)

U. S. Agriculture dept. Fire and the forest. 1925. 20 p. (United States Agriculture dept. Circ. 358)

—— —— Live stock grazing as a factor in fire protection on the national forests. 1920. 11 p. (United States Agriculture dept. Circ. 134)

—— —— Persons injured fighting forest fires. . . . 1912. 9 p. (62d Cong., Senate doc. 372)   Serial 6181

—— *Budget bureau.* Fires and floods in national parks, . . .
appropriation required for emergency reconstruction and
fighting forest fires in national parks. . . . 1926. 2 p.
(69th Cong., Senate doc. 59)   Serial 8557
—— —— Forest service, forest roads and trails, . . . supplemental estimates of appropriations, 1927, for fighting and
preventing forest fires, for forest roads and trails. . . . 1926.
4 p. (69th Cong., House doc. 600)   Serial 8734
—— —— National park service, fighting forest fires, estimate of appropriation for Department of interior, National
park service, 1927. 1926. 2 p. (69th Cong., House doc. 576)
                                            Serial 8734
—— —— Supplemental estimate of appropriation for Forest
service. . . . 1927. To provide additional fire protection in
forests in Southern California. 1926. 3 p. (69th Cong.,
House doc. 410)   Serial 8579
—— *Congress. House. Committee on agriculture.* Report to
accompany S. 3108, to amend sec. 2 of act, as amended (for
protection of forest lands, for reforestation of denuded areas,
for extension of national forests, and for other purposes, in
order to promote continuous production of timber on lands
chiefly suitable therefor, relating to protection from forest
fires). 1926. 3 p. (69th Cong., House rept. 777)
                                            Serial 8533
—— —— —— —— . . . Report to accompany S. 4224,
to amend Sec. 2 of act for protection of forest lands, . . .
relative to protection of certain watersheds from forest fires.
1925. 2 p. (68th Cong., House rept. 1582)   Serial 8391
—— —— —— —— Safeguarding national forests, hearings . . . on H. R. 4070, to provide coöperation to safeguard
endangered agricultural and municipal interests and to protect forest cover on Santa Barbara, Angeles, San Bernardino,
and Cleveland national forests from destruction by fire. . . .
Jan. 16 and 18, 1926. 1926. 91 p.   Serial D
—— —— —— —— To safeguard agricultural and municipal interests in Santa Barbara, Angeles, San Bernardino,
and Cleveland national forests from destruction by fire. . . .
1926. 3 p. (69th Cong., House rept. 459)   Serial 8532
—— —— —— *Committee on appropriations.* Fire protection in California, extracts from hearings. 1926. 22 p.
—— —— *Senate. Committee on agriculture and forestry.*
Fire protection in Santa Barbara, Angeles, San Bernardino
and Cleveland national forests, hearings, on S. 574, to pro-

17

vide coöperation to safeguard endangered agricultural and municipal interests and to protect forest cover on Santa Barbara, Angeles, San Bernardino and Cleveland national forests from destruction by fire, 1926. 68 p.

—— —— —— —— . . . Report to accompany S. 3108 to amend sec. 2 of act, as amended, for protection of forest lands, for reforestation of denuded areas, for extension of national forests, and for other purposes, in order to promote continuous production of timber on lands chiefly suitable therefor, relating to protection from forest fires. 1926. 3 p. (69th Cong., Senate rept. 320)   Serial 8524

—— —— —— —— Santa Barbara national forest. To safeguard agricultural and municipal interests in Santa Barbara, Angeles, San Bernardino, and Cleveland national forests, report to accompany S. 2646. . . . 1926. 3 p. (69th Cong., Senate rept. 368)   Serial 8524

—— *Forest service.* Fire prevention and control on national forests. 1913. 20 p.

—— —— Fire protection in District 1, for use of forest officers. 1915. 117 p.

—— —— Forest fire control, by John McLaren. 1925. 14 p. (United States Agriculture dept. Misc. circ. 44)

—— —— Same. 1926.

Also published in Journal of agricultural research, Nov. 15, 1925, v. 31: 923-28.

—— —— Forest fire prevention; handbook for school children of California. 1923. 24 p. (United States Agriculture dept. Misc. circ. 7)

Describes fire prevention and fire fighting by Forest service.

—— —— Forest fire protection by the states as described by representative men at Weeks law forest fire conference. 1914. 85 p.

Coöperation with Forest service.

—— —— Forest fire protection under the Weeks law in coöperation with States, by J. G. Peters. 1913. 145 p. (Forest service Circ. 205)

—— —— Forest fires in the intermountain region. 1924. 16 p. (United States Agriculture dept. Misc. circ. 19)

—— —— Methods and apparatus for the prevention and control of forest fires, by D. W. Adams. 1912. 27 p. (United States Agriculture dept. Forest service bulletin 113)

—— —— National forest fire protection plans. 1911. 8 p.

—— —— Protection of forests from fire, by H. S. Graves. 1910. 48 p. (Forest service bulletin 82)

—— —— Systematic fire protection in the California forests. 1914. 99 p.

What forest protection means. (In Outlook, Apr. 25, 1923, v. 133: 742-43)

Wilhelm, Donald. Lookout above the mountain. (In Saturday evening post, Aug. 21, 1920, v. 193, no. 8: 40-54)

Fire protection.

Winters, S. R. Forest fire patrol by airplane and radio. (In Aviation, Sept. 14, 1922, v. 13: 282-83)

Women and planes vs. forest fires. (In Literary digest, Nov. 3, 1923, v. 79: 44-48)

Communications

Adams, Bristow. Telephones and the forest. (In American telephone journal, Oct. 6, 1906, v. 14: 218-21)

Rangers use carrier pigeons. (In Official record of United States Department of agriculture, Feb. 15, 1922, v. 1: no. 7: 3)

Tillotson, M. R. Use of the heliograph. (In Scientific American supplement, Aug. 31, 1918, v. 86: 141)

U. S. Forest service. Handbook on construction and maintenance of the national forests' telephone system. 1925. 126 p.

Early editions are entitled Telephone construction and maintenance on the national forests.

Maps, Roads, and Landscaping

Cobbs, J. L. Road building in the national forests. (In American automobile association, Highways greenbook, 1920: 35-42)

Federal funds for forest roads. (In Official record of United States Department of agriculture, June 25, 1924, v. 3, no. 26: 8)

Forest roads funds apportioned. (In Official record of United States Department of agriculture, Feb. 1, 1922, v. 1, no. 5: 2)

Hatton, J. H. National forest road building. (In Producer, Nov. 1921, v. 3, no. 6 : 5-8)

Joint administration of federal road fund. (In Official record of United States Department of agriculture, July 12, 1922, v. 1, no. 28 : 15)

Forest service coöperates with Public roads bureau.

Morrell, Fred. Relation of roads to forest management. (In Journal of forestry, Nov. 1927, v. 25 : 818-34)

Road in the forest, necessary, but why? (In Outlook, Mar. 25, 1925, v. 139 : 444-46)

Criticism of Forest service.

Sherman, E. A. Forest service and preservation of natural beauty. (In Landscape architecture, Apr. 1916, v. 6, no. 3 : 115-19)

Trails constructed by national Forest service. (In Engineering news, Sept. 19, 1918, v. 81 : 539-40)

—— *Forest service*. Application for national forest road under section 8, Federal road act, approved July 11, 1916. 1916. 7 p.

(Form 70)

—— —— Landscape engineering in the national forests, by F. A. Waugh. 1918. 38 p.

—— —— Manual for forest development, roads and minor roads. 1926. 195 p.

Mimeographed.

—— —— Maps.

Besides the regular maps, the Forest service publishes folder maps, similar to railroad folders, with map on one side of large sheet and information on reverse side all folded to railroad timetable size. A list of these maps may be found on pages 119-22 of " List of publications of United States Department of agriculture, by M. G. Hunt. 1901-1925. 1927. 182 p."

—— —— Mount Hood loop highway log, Mount Hood national forest. 1926. 6 p.

—— —— National forest road law. . . . 1916. 4 p. (64th Cong., House doc. 1336)   Serial 7099

—— —— Preparation of the forest atlas. 1907. 4 p.

—— —— Trail construction on the national forests. 1915. 69 p.

—— —— Same. Rev. ed. 1923. 81 p.

—— *Public roads bureau.* Federal legislation providing for Federal aid in highway construction and construction of national forest roads and trails. . . . 1927. 30 p. (United States Agricultural dept. Misc. circ. 105)

—— —— General instructions to resident engineers and superintendents of construction on national forest roads and national parks roads. 1926. 33 p.

Wilderness forest of Minnesota will not be traversed by roads. (In Official record of United States Department of agriculture, Sept. 22, 1926, v. 5, no. 38:4)

RECREATION IN NATIONAL FORESTS

Chapman, H. H. Recreation as a federal land use. (In American forests and forest life, June 1925, v. 31:349-51, 378-80)

Greeley, W. B. Recreation in the national forests. (In American review of reviews, July 1924, v. 70:65-70)

Gregg, W. C. Has our forest service gone "daffy"? (In Outlook, Feb. 11, 1925, v. 139:226-27; also Discussion, in same, Mar. 4, 11, 1925, v. 139:336-39, 365-66)

Criticism of the Forest service for too much attention to recreation on the national forests.

Kneipp, L. F. Recreation value of national forests. (In Parks and recreation, Mar.-Apr. 1925, v. 8, no. 4:300-04)

Mayer, M. H. Forests or forestry. (In Journal of forestry, May 1925, v. 23:451-56)

Moon, F. Forest recreation. (In Playground, July 1924, v. 18:218-19)

Rhoades, Verne. Federal forest purchases and forest recreation. Chapel Hill, N. C. University of N. C. Geological and economic survey. 1924. 8 p.

Schreck, R. G. Recreation on the national forests. (In Ames forester, annual no. 1922, v. 10:15-22)

Scott, J. E. Forests as national playgrounds. (In American forests and forest life, Jan. 1925, v. 31:25-28, 52-54)

Stahl, C. J. Recreation policy of the Forest service. (In Trail and timberline, Denver, Colorado, Jan. 1920; no. 17:7-8)

U. S. Agriculture dept. Forest trails and highways of the Mt. Hood region, Oregon national forest. 1920. 32 p. (United States Agriculture dept. Circ. 105)

—— —— Handbook for campers on the national forests in
California. 1921. 48 p. (United States Agriculture dept.
Circ. 185)

—— —— In land of ancient cliff dweller, Bandelier national
monument, Santa Fe national forest. 1923. 18 p. (United
States Agriculture dept. Misc. circ. 5)

—— —— In the open: the national forests of Washington.
1920. 78 p. (United States Agriculture dept. Circ. 138)

—— —— Land of beautiful water. Chelan national forest.
1920. 14 p. (United States Agriculture dept. Circ. 91)

—— —— Mountain outings on the Rainier national forest.
1920. 26 p. (United States Agriculture dept. Circ. 103)

—— —— Mountain playgrounds of the Pike national forest.
1919. 17 p. (United States Agriculture dept. Circ. 41)

—— —— Out of door playgrounds of the San Isabel national
forest. 1919. 18 p. (United States Agriculture dept. Circ. 5)

—— —— Outdoor life in the Colorado national forest. 1919.
19 p. (United States Agriculture dept. Circ. 34)

—— —— Summer vacation in the Sopris national forest.
1919. 14 p. (United States Agriculture dept. Circ. 6)

—— —— Vacation land, the national forests in Oregon.
1919. 72 p. (United States Agriculture dept. Circ. 4)

—— —— Vacation trips in the Holy Cross national forest.
1919. 14 p. (United States Agriculture dept. Circ. 29)

—— —— Washington national forest; a mountain vacation
land. 1920. 10 p. (United States Agriculture dept. Circ.
132)

—— Forest service. Fishing, hunting, and camping on the
Cascade national forest. 1920. 31 p. (United States Agriculture dept. Circ. 104)

—— —— Ouray mountains of the Uncompahgre national
forest. 1919. 14 p.

—— —— Recreation uses on the national forests, by F. A.
Waugh. 1918. 43 p.

—— —— Vacation days in Battlement national forest.
1919. 13 p.

—— —— Vacation days in Colorado's national forests.
1919. 60 p.

—— —— Vacation land of lakes and woods; Superior national forest. 1919. 12 p.

—— —— Vacation trips in the Cochetopa national forest.
1919. 14 p.

Forest Ranger's Work

Allen, S. W. Rough-and-ready engineers. (In American forests
and forest life, Feb. 1925, v. 31: 67-70)

Ancona, E. P. Forest ranger receives a compliment. (In
Northwoods, Oct. 1921, v. 7, no. 28: 5-8)

Barnes, W. C. Diary of a forest ranger's wife. (In Saturday
evening post, Sept. 19, 1925, v. 198: 6-74)

—— Duties of forest rangers. (In Breeder's gazette, Mar.
18, 1915, v. 67: 558-59)

—— Forest service. (In Ames forester, annual no. 1927:
25-30)

—— When winter comes on the range. (In Field illustrated,
Feb. 1925, v. 35, no. 2: 11-13, 37, 50)

Bassett, R. O. Experiences of a forest guard in the United
States forestry service. (In Empire forester, 1916, v. 2, no.
1: 45-48)

Brill, E. C. Forestry on snowshoes. (In St. Nicholas, Feb.
1923, v. 50: 418-22)

Chapman, Arthur. Day with a forest ranger. (In Outlook,
Oct. 28, 1905, v. 81: 489-95)

—— On the telephone frontier. (In Technical world magazine, June 1914, v. 21: 583-85)

Chapman, H. H. Forest service and its men. (In Journal of
forestry, Oct. 1918, v. 16, no. 11: 653-70)

Cleator, F. W. Forest supervisor. (In Gopher Peavey forestry
club, University of Minnesota, 1922, v. 2: 30-31)

Cobbs, J. L. Uncle Sam's handy man [the forest ranger]. (In
Hunter-trader-trapper, Aug. 1917, v. 34, no. 5: 13-25)

Cope, J. A. A day with the Forest service. (In Outing magazine, Jan. 1916, v. 67: 405-08)

Crump, Irving. Boy's book of forest rangers. New York, Dodd,
Mead and company, 1924. 253 p.

DeCamp, A. S. Forest ranger's wife. (In American forestry,
Aug. 1923, v. 29: 47-73)

Elgin, M. Girard, a timber faring man of the mountains. (In
American forests and forest life, Jan. 1924, v. 30: 34-35,
51-52)

Ferguson, J. A. Position of forest assistant on the national
forests. (In Pennsylvania state farmer, May 1909, v. 2, no.
4: 86-88)

Frothingham, R. Expert in forestry. (In Sunset, June 1927,
v. 58: 47-48)

Hatton, J. H. Uncle Sam's buffaloes. (In Producer, Nov.-Dec. 1926, v. 8, no. 6-7: 3-6, 5-9)

Holt-Wheeler, F. W. Boy with U. S. foresters. . . . Boston, Lothrop, Lee and Shephard company, 1910. 317 p.

Jones, H. L. Forest ranger, his life, duties and responsibilities. (In Overland monthly, Nov. 1902, n. s. v. 40: 439-46)

Lawson, W. P. Log of a timber cruiser. New York, Duffield and company, 1915. 214 p.

Incidents of actual life and day-to-day duties of Service field men.

—— Uncle Sam's forest physicians. (In Harper's weekly, Sept. 4, 1915, v. 61: 230-32)

Training and duties of a forest assistant.

Leopold, Aldo. Forest service salaries and the future of the national forests. (In Journal of forestry, Apr. 1919, v. 17: 398-401)

McConnel, M. W. Why is a forest ranger's wife? (In American forestry, Jan. 1924, v. 30: 44-46)

McDonald, C. H. Two weeks with a forest ranger. (In American forests and forest life, Dec. 1927, v. 33: 720-21, 750)

McLaren, John. Answering the call. (In American forestry, Aug. 1922, v. 28: 471-72)

Munch, William. Every forest ranger and game warden a scout master. (In Northwoods, St. Paul, Feb. 1921, v. 7, no. 22: 13-20)

O'Brien, C. E. In a Forest service camp. (In Overland monthly, Feb. 1916, n. s. v. 67: 146-52)

Pinchot and the ranger schools. (In Outlook, Feb. 19, 1910, v. 94: 368-69)

Pinchot, Gifford. Spirit of the Forest service. (In California forestry, May 1917, v. 1, no. 1: 1)

Pittman, Alfred. Adventures of a forest ranger. (In American magazine, Aug. 1925, v. 100: 56-59, 94-98)

Pratt, L. C. Ranger's of the north. (In American forestry, Jan. 1924, v. 30: 20-22, 53)

Preston, J. F. Work of the forest rangers. (In Lumber, Aug 26, 1918, v. 62, no. 12: 18-26)

Redington, P. G. Forest service and the public. (In American forestry, July 1914, v. 20: 511-13)

Retirement of Alfred F. Potter. (In Journal of forestry, Mar. 13, 1920, v. 18: 211-13)

Commendation of his work as forest inspector, inspector of grazing and Assistant forester.

Ritchie, R. W. Soldiers of the pines. (In Country gentlemen, Jan. 10, 1925, v. 90: 5, 30)

Robertson, M. H. Forest ranger. (In Illustrated world, Mar. 1922, v. 37: 58-60)

Romance and real work in the Forest service. (In American lumberman, May 31, 1913, no. 1985: 56-57)

Rutledge, R. H. Duties of wardens and patrolmen. (In Timberman, Oct. 1919, v. 20, no. 12: 83-84)

Shinn, C. H. Work on a national forest. (In Conservation, May, July, Oct., Dec. 1907; March, May, July, Aug. 1908; April, July, Oct. 1909, v. 13: 240-43, 375-81, 520-24, 590-99, 637-47; v. 14: 166-69, 243-47, 383-86, 473-80; v. 15: 187-96, 397-401, 623-26)

Shinn, J. T. Clerk's work on a national forest. (In American forestry, May 1915, v. 21: 653-56)

Shipp, T. R. Forestry work that is being done for the preservation of the nation's timberland; scope of the task and story of the men who are saving the woods. (In Reader, July 1906, v. 8: 167-78)

Stratton, G. F. Rangerettes and rangers. . . . (In Country gentleman, Dec. 6, 1924, v. 89: 3, 26)

Tracy, R. P. Roughneck pioneers. (In Sunset, Feb. 1922, v. 48: 32-35, 66)

Tribute to the rangers. (In North woods, Jan. 1919, v. 7, no. 1: 16-18)

U. S. Agriculture dept. Farm forestry extension. 1925. 14 p. (United States Agriculture dept. Circ. 345)

—— Forest service. Information regarding employment on national forests. 1905-date.

Reissued from time to time with revisions.

Wygle, E. P. Ex-12; ex-guard, ex-ranger. (In Ames forester, (Ames, Iowa.) 1914. v. 2: 35-40)

FOREST CONSERVATION AND REFORESTATION

Adams, Bristow. What we can do to preserve our forests. An account of the work of the Forest service. . . . (In New Idea Women's magazine, Mar. 1908, v. 17: 28-30)

American reforestation association. Comprehensive nation-wide plan for conservation and reforestation . . . Los Angeles, American reforestation association, 1923. 46 p.

Ayres, P. W. Tackling the forestry problem in time. (In American Review of reviews, July 1922, v. 66 : 73-76)

Baker, H. P. Afforestation and paper making. (In Paper, Feb. 15, 1922, v. 29 : 13-14)

Point of view of paper manufacturers toward government's forest policy.

Beveridge, A. J. National Forest service. . . . Washington, no publisher, 1907. 33 p.

Calkins, M. C. Clark forestry law. (In Journal of land and public utility economics, Jan. 1925, v. 1 : 126-28)

Carter, C. F. Our reforestation activities. (In Scientific American, Dec. 1921, v. 125, A : 106-07)

Facts on the Clarke-McNary forestry act. (In Railway review, Apr. 18, 1925, v. 76 : 737)

Forestry and the federal budget. (In American forests and forest life, Nov. 1925, v. 31 : 668)

Fox, F. C. Conservation of forests. (In U. S. Bureau of education, 1927, v. 20 : 40-58)

Gilmore, C. L. Federal domination vs. state sovereignty. (In Mining congress journal, Mar. 1927, v. 13 : 179-82)

Graves, H. S. Federal and state responsibilities in forestry. (In American forests and forest life, Nov. 1925, v. 31 : 675-77)

—— Public forests and private forestry. (In American forestry, Oct. 1923, v. 29 : 613-14)

Greeley, W. B. Economic aspects of forestry. (In Journal of land and public utility economics, Apr. 1925, v. 1 : 129-37)

—— Federal government's policy for southern national forests. (In Southern lumberman, Jan. 16, 1926, v. 122, no. 1581 : 35-36)

—— Putting our idle forest acres to work. (In American Review of reviews, Feb. 1925, v. 71 : 189-92)

—— Shall the national forests be abolished? (In Mining congress journal, Aug. 1927, v. 13 : 594-97)

Guthrie, J. D. Forestry awakening in Washington. (In American forestry, Jan. 1922, v. 28 : 51).

Knapp, G. L. Other side of conservation. (In North American review, Apr. 1910, v. 191 : 465-81)

Knappen, T. M. Shall our forests be "developed" or renewed?
(In World's work, May 1922, v. 44, no. 1: 78-88)

Lake, L. N. Guarding our forest empire. (In Travel, Aug.
1925, v. 45: 11-15)

Larger appropriations urged for federal forestry. (In American forests and forest life, Nov. 1925, v. 31: 685)

Leaders of paper industry to aid in forestry policy [of conservation]. (In Official record of United States Department of
agriculture, Oct. 10, 1923, v. 2, no. 41: 3)

Michigan forestry commission. American Forest service [replanting California big trees]. (In Michigan forestry commission, Report, 1905-1906: 160-63)

Morrow, J. B. Planks for posterity; something of the mighty
task facing Col. Greeley. (In Nation's business, Nov. 1920,
v. 8, no. 11: 36-40)

Morse, H. B. Government responsibility in forest regulation.
(In Journal of forestry, Mar. 1923, v. 21: 269-72)

Mulford, W. The challenge. (In Journal of forestry, Nov.
1925, v. 23: 863-68)

Pack, A. N. Our vanishing forests. New York, Macmillan
company, 1923. 189 p.
Describes work of the government in forest preservation.

Pack, C. L. Forest production; words or actions! (In Nature
magazine, Jan. 1924, v. 3: 27-28)

—— The forestry primer, 1876, 1926. Washington, D. C.
American tree association, 1926. 32 p.

Page, A. W. Statesmanship of forestry, the government's
method of perpetuating our forests and what it means to the
future of the country. (In World's work, Jan. 1908, v. 15:
9739-58)

Page, W. H. Gifford Pinchot the awakener of the nation. (In
World's work, Mar. 1910, v. 19: 12662-68)

Palmer, Frederick. Pinchot's fight for the trees. . . . (In Colliers weekly, Nov. 30, 1907, v. 40, no. 10: 13-14)

Paper industry endorses forestry bill. (In Paper trade journal, Mar. 27, 1924, v. 78: 23-24)

Peters, J. G. Work of the Forest service in the South. (In
American lumberman, Oct. 15, 1910, no. 1847: 54-55)

Pinchot, Gifford. Menace to our forests. (In Scientific American, July 1922, v. 127: 52)

Pratt, G. D. Legislative needs in forestry. (In American forests and forest life, Oct. 1926, v. 32: 603-04)

Preston, J. F. and Eldredge, I. F. High spots of modern management plans for the national forests. (In Journal of forestry, Feb. 1923, v. 21: 116-24)

Price, O. W. Task of the Forest service. (In Independent, Sept. 2, 1909, v. 67: 537-45)

Redington, P. G. Coöperation with lumberman. (In Timberman, May, 1924, v. 25: 194-98)

Reynolds, H. A. Wanted: a Department of conservation. (In American forests and forest life, May 1924, v. 30: 259-60, 308, 314)

Senators stress evil of forest depletion. (In American forestry, Feb. 1924, v. 30: 92, 115-16)

Smith, H. A. Saving the forests. (In National geographic magazine, Aug. 1907, v. 18: 519-34)

Tillotson, C. R. Reforestation on the national forests. 1917. 63 p. (United States Agriculture dept. Bulletin 475. Professional paper)

U. S. Agriculture dept. Importance of forestry and the national forests. 1924. 16 p. (United States Agriculture dept. Misc. circ. 15)

—— —— Truckee-Carson-Lake Tahoe project. Letter from Secretary of agriculture, transmitting in response to House res. 270, information relating to Truckee-Carson-Lake Tahoe project. 1912. 160 p. (62d Cong., House doc. 451)

—— *Congress. House. Committee on agriculture.* Acquisition of forest lands, report to accompany H. R. 271 [for protection to watersheds of navigable streams and to appoint commission for acquisition of lands for purpose of conserving navigability of navigable rivers]. . . . 1926. 4 p. (69th Cong., House rept. 431)   Serial 6321

—— —— —— —— Authorizing Secretary of agriculture to coöperate with territories in promoting continuous production of timber, report to accompany S. J. Res. 37 [so as to provide for protection of forest lands, for reforestation of denuded areas, for extension of national forests and for other purposes]. 1926. 3 p. (69th Cong., House rept. 778)
                                                                       Serial 8533

—— —— —— —— McNary-Woodruff bill hearings . . . on H. R. 271 authorizing . . . any state to coöperate with any other state or states or with United States for protection of watersheds of navigable streams and to appoint commission for acquisition of lands for purpose of conserving navigability of navigable rivers. . . . 1926. 53 p.   Serial H

—— —— —— —— . . . Protection of forest lands. . . .
Report (to accompany H. R. 4830) 1924. 8 p. (68th Cong.,
House rept. 439) Serial 8228
—— —— —— —— Reforestation act, amendments. Hearings . . . Feb. 24, 1925. 1925. 41 p. Serial EE
—— —— —— —— Reforestation. Hearings . . . on H.
R. 4830, a bill to provide for the protection of forest lands,
for reforestation of denuded areas. . . . Mar. 25, 27, 1924.
1924. 98 p.
—— —— —— —— . . . Report to accompany H. R.
9039 [to enable any state to coöperate with any other state or
states or with the United States for protection of watersheds of
navigable streams and to appoint commission for acquisition
of lands for purpose of conserving navigability of navigable
rivers, relative to payment of lands acquired by condemnation proceedings] 1926. 2 p. (69th Cong., House rept. 460)
Serial 8532
—— —— —— Committee on rules. Ballinger reports.
1911. 29 p. (61st Cong., House rept. 2102) Serial 5852
—— —— Senate. Committee on agriculture and forestry.
Navigability of navigable rivers hearings on H. R. 9039, to
amend sec. 8 of act to enable any state to coöperate with any
other state or states or with United States for protection of
watersheds of navigable streams. . . . 1926. 2 p.
—— —— —— —— Protection and promotion of forest
lands, report to accompany S. J. Res. 37. . . . 1926. 2 p.
(69th Cong., Senate rept. 328) Serial 8524
—— —— —— —— Protection of watersheds of navigable
streams, and conserving navigability of navigable rivers, report to accompany S. 3663. . . . 1925. 3 p. (68th Cong.,
Senate rept. 1064) Serial 8388
—— —— —— —— Protection of watersheds of navigable
streams, hearings . . . on S. 718, authorizing appropriation
to be expended under provisions of sec. 7 of act to enable any
state to coöperate with any other state or states or with United
States for protection of watersheds of navigable streams, and
to appoint Commission for acquisition of lands for purpose
of conserving navigability of navigable rivers. . . . 1926.
59 p.
—— —— —— —— Protection of watersheds of navigable
streams, report to accompany H. R. 9039. . . . 1926. 2 p.
(69th Cong., Senate rept. 618)

—— —— —— —— Protection of watersheds of navigable streams, report to accompany S. 718. . . . 1926. 4 p. (69th Cong., Senate rept. 366) Serial 8525

—— —— —— —— Reforestation. Hearing. . . . on S. 1182, a bill to provide for the protection of forest lands, for the reforestation of denuded areas, for the extension of national forests and for other purposes. . . . Mar. 28, 1924. 1924. 67 p.

—— —— —— *Select committee on reforestation.* Reforestation. Hearings. . . . pursuant to S. Res. 398 . . . Mar. 7, 1923–Nov. 23, 1923. 1923-24. 8 pts.

—— *Forest service.* Present needs in national and state forestry, by W. B. Greeley. 1927. 8 p.

—— —— Reforestation on the national forests, by W. T. Cox. 1911. 57 p. (United States Agriculture dept. Forest service bulletin 98)

—— —— Tree distribution under the Kincaid act of 1911. 1925. 14 p. (United States Agriculture dept. misc. circ. 16)

—— ——Work of the Division of forestry for the farmer, by Gifford Pinchot. 1898. (United States Agriculture dept. Yearbook, 1898: 297-308)

—— *General land office.* Instructions for making final proof on timber culture entries under act of Congress to encourage the growth of timber on western prairies. 1882. 12 p.

—— —— Instructions to special agents of General land office, appointed to prevent timber depredations and to protect public timber from waste and destruction. 1883. 39 p.

—— —— Preventing depredations on public timber . . . recommending appropriation of one million dollars for preventing depredations on public timber. 1909. 12 p. (60th Cong., Senate doc. 667) Serial 5408

—— *President* (Harrison) Message . . . transmitting report relative to preservation of forests on the public domain. 1890. 4 p. (51st Cong., 1st sess. Senate Ex. doc. 36) Serial 2682

—— —— (Roosevelt) Message . . . read at the 14th irrigation Congress held at Boise, Idaho, from Sept. 3-8, . . . 1906. 4 p.

Published also in Forestry and irrigation, Sept. 1905, v. 12: 399-401.

Unspoiled country needs protection. (In Orange Judd Illinois farmer, Oct. 15, 1926, v. 74: 612)

Walker, J. B. Uncle Sam spendthrift. (In Scientific American, Apr. 1926, v. 134: 230-32)

Wallace, H. C. Our forest problem. (In California cultivator, Jan. 13, 1923. v. 60:58)

White, S. E. Fight for the forests. . . . (In American magazine, Jan. 1908, v. 65: 252-61)

—— Woodsmen spare these trees. (In Sunset, Mar. 1920, v. 44, no. 3: 23-26)

Methods of Forest service in dealing with forest insects.

Winslow, C. P. Efficient forest utilization—a major factor in an effective national forest policy. (In Paper trade journal, Mar. 13, 1924, v. 78:30-34)

Winters, S. R. Reforesting denuded areas. (In Field illustrated, Nov. 1923, v. 33:12-14, 48)

Woehlke, W. V. Real forestry at last. (In Sunset, Apr. 1922, v. 48:16-19)

BIBLIOGRAPHY

Burns, Findley. Publications of the Forest service, classified for teachers and students. (In School science and mathematics, Nov. 1918, v. 18: 716-22)

Journal of forestry. 1917-date.

Monthly numbers of journal of forestry contain notices of new books on forestry.
This journal started as Forestry quarterly in 1902.

U. S. Agriculture dept. Catalogue of publications relating to forestry in the library of United States Department of agriculture, 1912. 302 p.

—— —— General index to the agricultural reports of the Patent Office for twenty-five years from 1837-1861 and of the Department of agriculture for fifteen years from 1862-1876. 1879. 225 p.

—— —— Supplement to General index . . . for years 1877-1885, inclusive. 1886. 113 p.

—— —— Index to annual reports of the Department of agriculture. . . . 1837-1893, inclusive. 1896. 252 p.

—— —— Index to yearbook of U. S. Department of agriculture, 1894-1900. 1902. 196 p.

—— —— Same. 1901-1905. 166 p.

—— —— Same. 1906-1910. 146 p.

—— —— Same. 1911-1915. 1915. 178 p

See also indexes in each annual volume.

——— ——— List by titles of publications of the U. S. Department of agriculture, 1840-1901, inclusive. . . . 1902. 216 p.

——— ——— List of publications of the Agriculture department, 1862-1902. 1904. 623 p.

——— ——— List of publications . . . from Jan. 1901 to Dec. 1925, inclusive, by M. G. Hunt. 1927. 182 p.

Forest service, p. 114-125.

——— ——— List of publications issued since July 1, 1913. Revised to Dec. 31, 1916. 1917. 114 p.

——— ——— List of publications of Department of agriculture for the five years 1889-1893, inclusive. . . . 1894. 42 p.

——— ——— Monthly list of publications, Mar. 1896-1924. 1896-1924.

Continued by monthly list printed on postal card and mailed to subscribers.

——— ——— Official record, ... Weekly. 1922-date.

The Official record of the Department of agriculture gives the latest developments in the work of all services and bureaus, as well as lists of the most important articles in outside periodicals concerning work of each Bureau.

——— ——— Weekly newsletter, Aug. 13, 1913–Dec. 7, 1921. 1913-1921.

Continued by Official record.

——— ——— *Division of publications.* Forest service. 1911-1913. (United States Agriculture dept. Division of publications, Circ. no. 11)

——— *Forest service.* Classified list of publications available for distribution, April 1, 1909. 1909. 4 p.

——— ——— Complete list of Forest service publications. 1915. 40 p.

——— ——— Same. 1925. 57 p.

——— ——— Same. Supplementary list, 1915-1921. 13 leaves.

The last two mimeographed.

——— ——— Some books on forestry. 1924. 11 p.

Mimeographed and in library of Forest service, Washington, D. C.

——— Superintendent of documents. Forestry. . . . 1930. 17 p.

(Price list 43—23d ed.)

INDEX

18   261

National Capital Park and Planning Commission, Forester a member of, 122.

National Committee on Wood Utilization, establishment of, 64.

National Conservation Commission, 36.

National Forest Reservation Commission, organization and function, 42, 90, 91, 122.

National Forests (*See also Appalachian Region; Forest Reserves; White Mountains*)

Agricultural claims, 54.

Area, 119.

Camping facilities, 63, 79.

Collections from, 35.

Creation and enlargement restricted in certain states, 35.

Designation changed from "forest reserves" (1907), 35.

District No. 7, created (1914), 56.

District No. 8, created (1921), 56.

District No. 9, created (1929), 56.

District organization, 112, 116.

Game preserves, 156.

Forest Homestead Act, 1906. See *Acts Cited*.

Homestead settlers, 35, 55.

Insect ravages, 57, 80, 82, 83.

Monuments in

Acreage of, 35, 85.

Bandelier Monument, 96.

List of, 120.

Recreational facilities, 54.

Revenues, disposition of, 35.

Roads and Trails

Forest Service transfers to road and trail fund, 211.

Road program, 52, 57, 86.

Timber removal, 74.

Timber sales, 72.

Timber settlement, 74.

Timberland area, 71.

Tree Diseases

Blister rust host eradication, 89.

Control methods, 84.

Losses from, 80.

Use Book, 75.

Utilization revenue, 79.

Water power. See *Power development*.

Working circles, 69.

Navy. See *Ship-construction Timber*.

Nebraska, "Arbor Day," originated in, 5.

New York State

Forest preserves, origin of, 16.

State ownership of wild land commission (1866), 5.

North American Conservation Conference, 36.

"Olympic blowdown," 81.

Paddock, Senator, of Nebraska, bills introduced, 20.

Patent Office, forest trees of North America, report (1860), 5.

Philippine Forest Bureau, relationship with Agriculture Department, 28.

Pinchot, Gifford

Chairman, National Conservation Commission, 49.

Decentralization policy, 37.

Dismissal (1910), 39.

Head of Division of Forestry, Department of Agriculture, 27.

Plant Industry Bureau, tree disease control work, 84.

Power Development in National Forests

Legislation, 46-52.

Regulations of 1910, 50.

Pinchot policy, 47.

United States v. Utah Light and Power Co., 51.

Water power permits, receipts from, 79.

President Arthur, messages on Indian land depredations, 17.

President Cleveland

Messages on Indian land depredations, 17.

Proclamations (1897), 21.

President Grant, special messages, (1874), 6.

President Harrison, message (1890), Indian land depredations, 17.